Numerical Solution of Partial Differential Equations

NATO ADVANCED STUDY INSTITUTES SERIES

Proceedings of the Advanced Study Institute Programme, which aims at the dissemination of advanced knowledge and the formation of contacts among scientists from different countries

The series is published by an international board of publishers in conjunction with NATO Scientific Affairs Division

A	Life Sciences	Plenum Publishing Corporation
B	Physics	London and New York
C	Mathematical and Physical Sciences	D. Reidel Publishing Company Dordrecht and Boston
D	Behavioral and Social Sciences	Sijthoff International Publishing Company Leiden
E	Applied Sciences	Noordhoff International Publishing Leiden

Series C – Mathematical and Physical Sciences

Volume 2 – Numerical Solution of Partial Differential Equations

Numerical Solution of Partial Differential Equations

Proceedings of the NATO Advanced Study Institute held at Kjeller, Norway, August 20–24, 1973

edited by

J. G. GRAM

Institutt for Atomenergi, Kjeller, Norway

D. Reidel Publishing Company

Dordrecht-Holland / Boston-U.S.A.

Published in cooperation with NATO Scientific Affairs Division

First printing: December 1973

Library of Congress Catalog Card Number 73–91204

ISBN 90 277 0413 9

Published by D. Reidel Publishing Company
P.O. Box 17, Dordrecht, Holland

Sold and distributed in the U.S.A., Canada, and Mexico
by D. Reidel Publishing Company, Inc.
306 Dartmouth Street, Boston, Mass. 02116, U.S.A.

Printed in The Netherlands by D. Reidel, Dordrecht

CONTENTS

PREFACE

This book contains the transcripts of the invited lectures presented at the NATO Advanced Study Institute on "Numerical Solution of Partial Differential Equations". The Study Institute was held at the Netherlands-Norwegian Reactor School, Institutt for Atomenergi, Kjeller, Norway, 20th - 24th August 1973. The members of the Scientific Advisory Committee were:

A.R. Mitchell, University of Dundee, Scotland
I. Holand, University of Trondheim, Norway
T. Håvie, University of Trondheim, Norway

The members of the Organizing Committee were:

E. Andersen, Institutt for Atomenergi, Kjeller, Norway
G.E. Fladmark, Institutt for Atomenergi, Kjeller, Norway
J.G. Gram, Institutt for Atomenergi, Kjeller, Norway

The aim of the Study Institute was to bring together mathematicians and engineers working with numerical methods. The papers presented covered both theory and application of methods for solution of partial differential equations. The topics were finite element methods, finite difference methods, and methods for solution of linear and nonlinear systems of equations with application to continuum mechanics and heat transfer.

The total number of participants was 68. Their names are given at the end of the book. The publication of these proceedings could be realized through the kind cooperation of the lecturers. The Advanced Study Institute was financially sponsored by NATO Scientific Affairs Division. The Organizing Committee wishes to express its gratitude for this support. Valuable assistance was given by Mrs. G. Jarrett who took care of many of the practical arrangements during the meeting (like hiring airplanes for participants wanting to go sightseeing in the Oslo-area). We also wish to thank our colleagues at Institutt for Atomenergi for their help in arranging the Study Institute.

Kjeller, September 1973. The Organizing Committee

METHODS FOR SOLUTION OF PARTIAL DIFFERENTIAL EQUATIONS

L. Collatz

Universität Hamburg,
Hamburg, Germany

SUMMARY

In this survey we do not intend to mention all types of numerical procedures but to look only on some special methods, which were used frequently in recent years or which deserved perhaps to be used more than hitherto.

I. DISCRETIZATION METHODS

1. General Formulation

Let us formulate the problems not in greatest possible generality, but so general, that many applications are contained in this formulation.

Let B be a domain in the n-dimensional point space R^n with coordinates $x_1,\ldots,x_n$ and ∂B the boundary of B. $u(x_j)$ may be an unknown function or a vector with m components $u_{(1)}(x_j),\ldots,u_{(m)}(x_j)$; the linear or nonlinear differential equation

$$Mu = 0 \text{ in } B \tag{1.1}$$

and the boundary conditions

$$Su = 0 \text{ on } \partial B \tag{1.2}$$

may be prescribed, M and S may be vectors. In problems with free boundaries B is not given.

There are many different kinds of discretization. We mention four methods

Finite difference methods
Finite differences of higher approximation
Hermitean methods ("Mehrstellenverfahren")
Finite element methods.

In all cases we have the question of

consistency
convergence
stability

of the discretization methods.

2. Finite Difference Methods

It is not necessary to explain this very often used and very well known method (compare Isaacson-Keller (66) Mitchell (69) a.o.). Only for comparing with the other methods we consider the classical case of Laplaces equation for a function u(x,y)

$$\Delta u = \frac{\partial^2 u}{\partial x^2} + \frac{\partial^2 u}{\partial y^2} = 0 \tag{2.1}$$

in the square grid with meshsize h

$$x_j = x_o + jh,\ y_k = y_o + kh\ (j,k = 0,\pm 1,\pm 2,\ldots) \tag{2.2}$$

and we have in the well known form of patterns the formula

$$\frac{1}{h^2}\begin{pmatrix} & 1 & \\ 1 & -4 & 1 \\ & 1 & \end{pmatrix} u - \begin{pmatrix} & 0 & \\ 0 & 1 & 0 \\ & 0 & \end{pmatrix} \Delta u = O(h^2)\ ; \tag{2.3}$$

this method was used for treating complicated problems; with the aid of solving large systems of linear or nonlinear equations, for boundary and initial value problems; (Varga (62) Ortega-Rheinboldt (70) a.o.). The many improved methods for solving big systems of equations are well known. A further method was discussed by Young (73).

In the following simple example we will compare discretization methods with parametric methods.

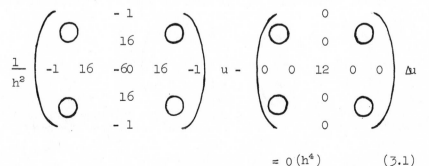

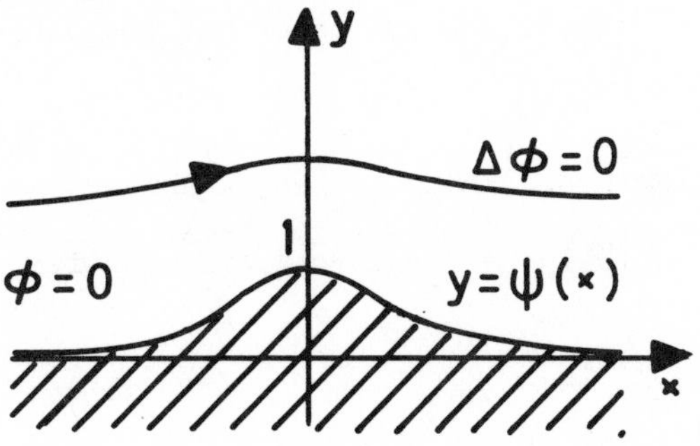

Fig. 1.

We consider the ideal flow of a liquid over a sill, fig. 1. The streamlines are given by Φ = const, where $\Phi(x,y)$ satisfies the conditions

$$\Delta\Phi = 0 \text{ in } B \; \left(-\infty < x < \infty,\; \Psi(x) < y < \infty,\; \Psi(x) = \frac{1}{1+x^2}\right) \tag{2.4}$$

$$\Phi = 0 \text{ for } y = \Psi(x)$$
$$\lim_{y\to\infty} [\Phi(x,y)-y] = 0 \text{ for every fixed } x. \tag{2.5}$$

The calculation of Φ by discretization methods causes in my opinion more work than by parametric methods in No. 10.

3. Finite Differences Of Higher Approximation

Here one is using other patterns for which the remainder term is of higher order. For the case (2.1) (2.2) we have for instance the formula

$$\frac{1}{h^2}\begin{pmatrix} & & -1 & & \\ & \bigcirc & 16 & \bigcirc & \\ -1 & 16 & -60 & 16 & -1 \\ & \bigcirc & 16 & \bigcirc & \\ & & -1 & & \end{pmatrix} u - \begin{pmatrix} & & 0 & & \\ & \bigcirc & 0 & \bigcirc & \\ 0 & 0 & 12 & 0 & 0 \\ & \bigcirc & 0 & \bigcirc & \\ & & 0 & & \end{pmatrix} \Delta u$$

$$= O(h^4) \tag{3.1}$$

The formulas of this type have often the disadvantage of difficulties in the neighbourhood of the boundaries (Collatz (60)).

4. Hermitean Methods ("Mehrstellenverfahren")

Instead of (2.1) we consider a differential equation of the form

$$Lu = f(x_j, u) \text{ in } B \tag{4.1}$$

where L is a linear differential operator and the given linear or nonlinear function does not depend on partial derivatives of u.

One chooses in B and ∂B a finite set of points $P_1, \ldots P_q$, often as gridpoints of a regular grid and writes down for each of these points an equation of the form

$$\Phi = \sum_{\rho=1}^{q} [a_\rho u(P_\rho) + b_\rho Lu(P_\rho) + \sum_{\sigma=1}^{s} c_{\rho\sigma} L_\sigma Lu(P_\rho)] \tag{4.2}$$

Usually in the equation for the point P_k the sum contains non zero terms only for the values of ρ for which P_ρ is in the neighbourhood of P_k. The L_σ are fixed chosen differential operators, for instance $\frac{\partial}{\partial x_1}$, $\Delta, \ldots$. In the simplest case one puts $c_{\rho\sigma} = 0$ (compare Albrecht (57)(62), Collatz (60)(72)).

The values $a_\rho, b_\rho, c_{\rho\sigma}$ are to be determined in such a way that one gets a remainder term of an order in h as high as possible if one develops Φ by Taylor expansion with respect to u and the partial derivatives of u at P_k. Then we substitute Lu by f, $u(P_\rho)$ by approximate values $U(P_\rho)$, put $\Phi=0$ and have one of the equations for the $U(P_\sigma)$.

For illustration: We have for the case (2.1)(2.2) the formula

$$\frac{1}{h^2}\begin{pmatrix} -2 & -8 & -2 \\ -8 & 40 & -8 \\ -2 & -8 & -2 \end{pmatrix} u + \begin{pmatrix} 0 & 1 & 0 \\ 1 & 8 & 1 \\ 0 & 1 & 0 \end{pmatrix} \Delta u = O(h^4) \tag{4.3}$$

For parabolic equations Wirz (72) has got good numerical results. Hermitean methods for initial value problems see Collatz (72).

5. The Finite Element Method (F.E.M.)

We suppose that there exists a variational principle for the problem (1.1)(1.2) in the form, that we look for an extremum of a functional

$$\Phi = \int_B F\left(x_j, u, \frac{\partial u}{\partial x_k}, \frac{\partial^2 u}{\partial x_k \partial x_l}, \ldots, \frac{\partial^m u}{\partial x_{k_1} \ldots \partial x_{k_m}}\right) dx_j + \Psi \tag{5.1}$$

where corresponding to the special case Ψ may be zero or an integral over parts of the boundary ∂B or finite terms a.o. We substitute u by a function $w = w(x_j, a_\nu)$ depending on parameters $a_1, \ldots, a_p$ and get a system of (linear or nonlinear) equations for the a_ν (Ritz-procedure)

$$\frac{\partial \Phi}{\partial a_k} = 0 \quad (k = 1, \ldots, p) \tag{5.2}$$

Let be $\hat{a}_\nu$ a solution of (5.2), then we take $w(x_j, \hat{a}_\nu)$ as approximate solution of the problem.

In the F.E.M. we divide B into a finite number of subdomains or "elements" $B_1, \ldots B_s$, which are usually

in the plane R^2 triangles or rectangles or elements with algebraic curves as boundaries, fig. 2.

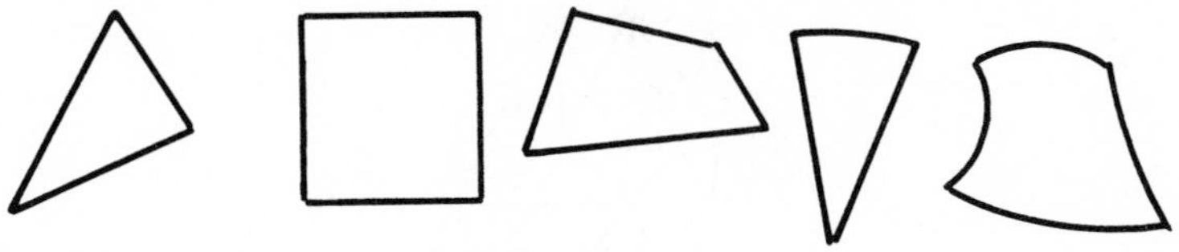

Figure 2.

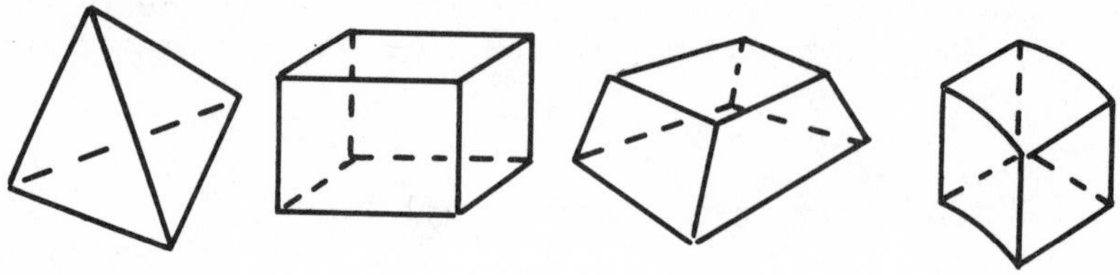

Figure 3.

in the 3-dimensional space R^3 tetrahedra a.o., fig. 3.

About curved shaped elements compare Mitchell-McLeod (73) and Wachspress (73).

In each element B_σ one takes for w as shape function usually polynomials, for instance splines (Strang (73) a.o.), sometimes also other functions. Fig. 4 gives an example for triangulation, Zienkiewicz (73).

At the interfaces one has inter-element conditions; if one wishes to have continuity up to derivatives of order (m-1), one has a conforming method, otherwise a nonconforming method, for instance Ciarlet (73). Special conditions may occur at the boundary.

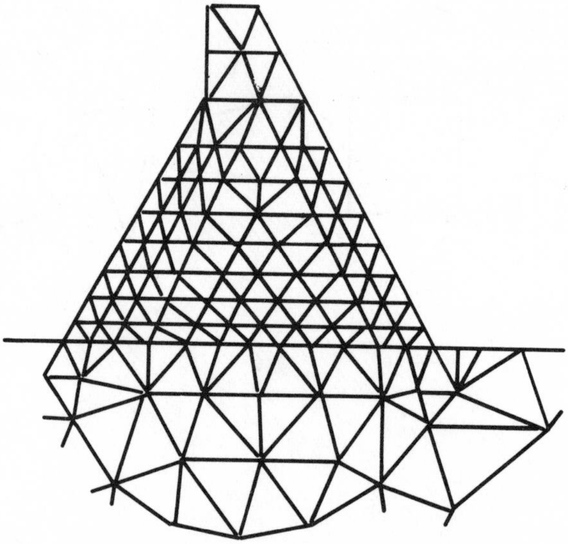

Figure 4

The F.E.M. is today the most used method for more complicated problems in continuum mechanics and therefore so well known, that it is not necessary to describe the method in more detail; compare Mitchell (73), Whiteman (73), Zienkiewicz (71)(73), Strang (73) a.o.

6. Comparison Of The Methods No. 2 To No. 5

The F.E.M. is preferred very often by engineers compared with No. 2 to No. 4:

a) the formulation (5.1) corresponds often physical quantities, energy a.o. and the method is therefore easily applicable

b) curved boundaries cause for methods No. 2 to 4 often considerably more difficulties than for F.E.M.

c) boundary conditions (for instance free boundary of a plate) are easier to handle

d) often one gets approximate values for derivatives (velocities, stresses, ...) easier

e) the division of B in elements B_σ can be done with more flexibility with respect to the problem.

On the other side: If one wishes to use formulas of higher accuracy one has to take polynomials of high degree in every element with many parameters, and this causes a great amount of work. For instance working with continuous shape functions one uses for the Laplace equation $\Delta u = 0$ in the plane the pattern (Heise (73))

$$\begin{pmatrix} -1 & -1 & -1 \\ -1 & 8 & -1 \\ -1 & -1 & -1 \end{pmatrix} U = 0 \qquad (6.1)$$

which is slightly more complicated as (2.3) and has no higher order of convergence as (2.3), but (4.3) has the same type as (6.1) with higher order of accuracy. For rectangles of course one can use F.E.M. with splines of C^1 (which have continuous derivatives), but then one has in every gridpoint four unknowns, the approximate values for $u, \frac{\partial u}{\partial x}, \frac{\partial u}{\partial y}, \frac{\partial^2 u}{\partial x \partial y}$. This gives the same order of accuracy as (4.3) but with nearly four times as many unknowns; however, one has also approximate values for the derivatives, which may be useful for applications.

In my opinion one has today not enough numerical experience for the comparison of the methods of No. 4, 5 and 10, particularly for nonlinear problems and using of approximation methods.

For all discretization methods exist practicable exact error bounds only in special cases. In the following we discuss certain possibilities for getting error bounds.

II. PARAMETRIC METHODS

7. Iteration Procedures

Often one can write the problem (1.1)(1.2) in the form

$$U = Tu \tag{7.1}$$

with a certain operator T (in many cases as an integral operator). Then one can try to use the iteration procedure

$$u_{n+1} = Tu_n \qquad (n = 0,1,2,\dots) \tag{7.2}$$

starting with anelement u_o of the domain of definition D of the operator T.

Let D be a domain in a partially ordered Banachspace R and $T : D \rightarrow R$.

The ordering may be written as $v < w$ or in special cases as $v \leq w$, if this means that $v(x) \leq w(x)$ in the classical sense for all $x_j \in B$.

Suppose T as syntone in D (Collatz (66)), (here No. 9) then one can ask for an element $u_o \in D$ with

$$0 \leq u_1 - u_o \leq \delta, \ \delta = \text{Min}. \tag{7.3}$$

This is a onesided Tschebyscheff Approximation (T.A.), written as optimization problem, Collatz-Wetterling (71), Laurent (72).

If one chooses u_o as element of a p-parametric family of functions

$$u_o \in W = \{w(x_j, a_1, \dots, a_p)\} \subset D \tag{7.4}$$

we have $u_o = u_o(x_j, a)$ and $u_1 = u_1(x_j, a_j)$; then (7.3) is a chained approximation; we call an approximation problem a chained one, if in the analytic expression (here $u_1 - u_o$) at least one parameter a_ν occurs at at least two places and if one cannot avoid this by a one to one transformation of the parameters.

8. Other Examples For Chained Approximation

Chained approximation occurs very frequently. Examples are for instance inverse problems in differential equations. Suppose we have observed a vibration y(t) satisfying a differential equation of the form (with unknown constants c, d)

$$\frac{d^2y}{dt^2} + cy = d$$

The general solution is

$$y(t) = \frac{d}{c} + a_1 \cos (\sqrt{c}\, t + a_2).$$

We consider y(t) as known from observations and wish to determine the physical constants c, d. Here c occurs twice, therefore we have a chained approximation.

Many other examples of chained approximation see Collatz (73).

9. Monotone Operators

There exists a theory on a general class of operators.

We consider operator T (D as domain of definition), working in partially ordered spaces.

If $v < w$ has the consequence $Tv < Tw$ for all $v,w \in D$, then T is syntone.

If $v < w$ has the consequence $Tv > Tw$ for all $v,w \in D$, then T is antitone.

<u>Theorem</u> (J. Schrøder (59)). Suppose $T = T_1+T_2$ with T_1 syntone, T_2 antitone, T_1 and T_2 compact operators, defined in a convex domain D. The iteration procedure

$$\begin{aligned} v_{n+1} &= T_1 v_n + T_2 w_n \\ w_{n+1} &= T_1 w_n + T_2 v_n \end{aligned} \qquad n = 0,1,2,\ldots \qquad (9.1)$$

may start with elements $v_o, w_o \in D$ with

$$v_o < v_1 < w_1 < w_o \tag{9.2}$$

Then exists at least one solution $u = Tu$ with $v_1 < u < w_1$.

Many applications to differential and integral equations were made, Collatz (66), Walter (70), an example here is given in No. 10.

Monotonicity principles hold for wide classes of linear and nonlinear elliptic and parabolic equations and in special cases also for hyperbolic equations, Collatz (66).

The Hammerstein-Integral equation

$$u(x_j) = \int_B K(x_j,t_k)\ \varphi\ (u(t_k))\ dt_k \tag{9.3}$$

can be treated in this way and therefore for instance the problem

$$\begin{aligned} -\ \Delta u &= f(u) \text{ in } B \\ u &= g(x) \text{ on } \partial B, \end{aligned} \tag{9.4}$$

if the given realvalued function $f(z)$ is of bounded variation.

Then one can write

$$f(u) = f_1(u) + f_2(u) \tag{9.5}$$

where f_1 is monotonically non decreasing and f_2 monotonically non increasing. The iteration procedure starts with two functions v_o, w_o which satisfy the boundary conditions, and v_1, w_1 satisfy

$$\left.\begin{aligned} -\ \Delta v_1 &= f_1(v_o) + f_2(w_o) \\ -\ \Delta w_1 &= f_1(w_o) + f_2(v_o) \end{aligned}\right\} \quad \text{in } B$$

$$v_1 = w_1 = g(x) \qquad \text{on } \partial B \tag{9.6}$$

10. Approximation And Combi-Approximation

In more complicated cases usually one cannot find functions which satisfy (1.1) or (1.2) exactly and then one comes to more general Combi-Approximations. One has then to approximate the differential equation in B and the boundary conditions on ∂B simultaneously (compare Bredendiek (70), Collatz-Krabs (73)).

If monotonicity principles are applicable, usually one gets exact error bounds for the solution. If no monotonicity properties hold, one does not get error bounds, but still the approximation methods are applicable to get approximate solutions. Usually as methods discretization, collocation and related methods are used.

In case of singularities (reentering corners, infinite domains, singularities of the coefficients a.o.) one can often use functions of the same type of singularities the solution has.

For applications to nonlinear vibrations see J. Werner (70).

<u>Example 1</u>: In the problem (2.4)(2.5), fig. 5, we put $u = y - \Phi$ and have

$$\Delta u = 0 \text{ in } B \left(-\infty < x < \infty,\ \Psi(x) = \frac{1}{1+x^2} < y < \infty\right)$$

$$u = y \text{ for } y = \Psi(x),\ \lim_{y \to \infty} u(x,y) = 0$$

for each fixed x.

We take as approximate solution $u \approx w(x,y) = \sum_{\nu=1}^{p} a_\nu w_\nu(x,y)$

with $\Delta w_\nu = 0$, for instance $w_\nu = \dfrac{y+c_\nu}{x^2+(y+c_\nu)^2}$.

If then holds $|w(x,\Psi(x)) - \Psi(x)| \leq \delta_p$ for $-\infty < x < \infty$, one has the exact error bound $|u-w| \leq \delta_p$ in B. Therefore one tries to determine the a_ν and c_ν in such a way that $\delta_p \approx$ Min. With only one term ($p=1$, $c_1=0$) one gets $\delta_1=0.17$; with two terms ($p=2$, $c_1=0$) and $a_1 = 0.79807$, $a_2 = 1.02523$, $c_2 = 6.3485$ one has the error bound $\delta_2 = 0.0624$. One can easily improve these bounds. (Putting $c_2 = 1$ one has simply a linear approximation problem with $\delta_2 = 0.0707$). I thank Mr. Heinz Günther and Mrs. Stefanie Moldenhauer for the numerical computation on a computer.

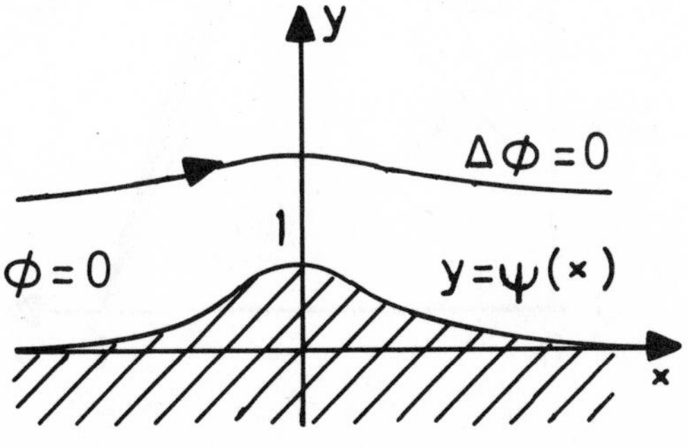

Fig. 5.

Example 2: We consider the nonlinear boundary value problem

$$-\Delta u = \frac{2+u}{2-u} \text{ in } B \; (r^2 = x^2 + y^2 < 1)$$

$$u = 0 \quad \text{on } \partial B \; (r = 1).$$

We iterate with

$$-\Delta v_1 = \frac{2+v_o}{2-v_o} \quad \text{in } B, \; v_o = v_1 = 0 \text{ on } \partial B$$

and take the simplest case with only one parameter a

$$v_o = (1-r^2)\,\frac{2(1-4a)}{-1-4a+(4a-1)r^2}$$

$$v_1 = \frac{(1-r^2)}{16}\;[12a+1-(4a-1)r^2]$$

With $0 \le v_1 - v_o \le \delta$, $\delta = \text{Min}$, we get a $a = 9/28$; analogeously we substitute v_o, v_1^o, a by w_o, w_1, b and can satisfy (9.2), fig. 6.

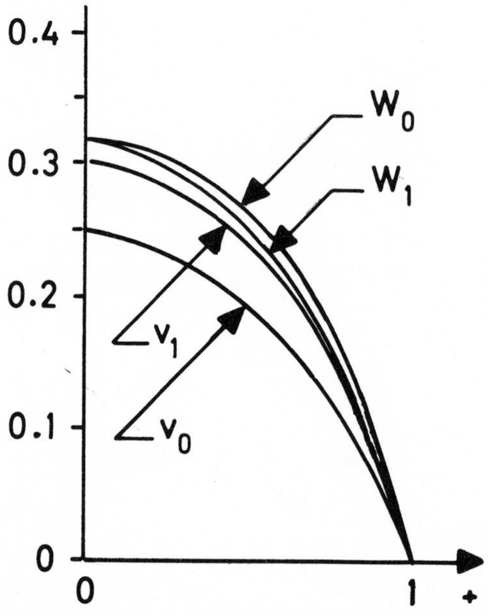

Fig. 6.

We get the exact error bounds (with $b = 1/12\ (14-\sqrt{97})$)

$$\frac{(1-r^2)(17-r^2)}{56} \leq u \leq \frac{1-r^2}{16}\ [12b+1-(4b-1)r^2],$$ for instance

$0.3035 < 17/56 \leq u(0,0) \leq 0.322.$

11. Other Methods

Very general applicable principles are variational principles, for instance for the problem (1.1)(1.2) the least square method

$$Q[w] = p(x_j) \int_B (Mw)^2 dx + q(x_j) \int_{\partial B} (Sw)^2 d\tau = \text{Min}$$

with given positive weight-functions p,q.

For linear selfadjoint (and certain other) problems functionals Q[w] can be used, No. 5, compare Ritz-Method, Galerkins method, Trefftz-Method a.o.

References

1. J. Albrecht (57), Zum Differenzenverfahren bei parabolischen Differentialgleichungen, Z. angew. Math. Mech. 37(1957), 202-212.

2. J. Albrecht (62), Zum Mehrstellenverfahren bei Kugel- und Zylindersymmetrie, Z. angew. Math. Mech. 42(1962), 397-402.

3. E. Bredendiek (70), Charakterisierung und Eindeutigkeit bei simultanen Approximationen, Z. angew. Math. Mech. 50(1970), 403-410.

4. P.G. Ciarlet (73), Conforming and nonconforming finite element methods for solving the plate problem, to appear Proc. Symp. Numerical solution of differential equations, Dundee 1973.

5. L. Collatz (60), The numerical treatment of differential equations, Springer 1960, 568 p.

6. L. Collatz (66), Functional Analysis and Numerical Mathematics, Aca. Press 1966, 473 p.

7. L. Collatz (70), Einseitige Tschebyscheff-Approximation bei Randwertaufgaben, Proc. Internat. Conference on Constructive funct. Theory, Varna, Bulgaria, 1970, 151-162.

8. L. Collatz-W. Wetterling (71), Optimierungsaufgaben, Springer Verlag, 2. ed., 1971, 222 p.

9. L. Collatz (72), Hermitean methods for initial value problems in partial differential equations, to appear Proc. Symp. Numer Methods, Dublin, Eire, 1972.

10. L. Collatz (73), Chained Approximation, to appear Proc. Symp. Funct. Anal. and Applicat., Madras, India, 1973.

11. L. Collatz-W. Krabs (73), Tschebyscheff Approximation, Teubner, 1973, to appear.

12. U. Heise (73), Combined application of finite element methods and Richardson extrapolation to the torsion problem, in Whiteman (73), 225-237.

13. E. Isaacson-H.B. Keller (66), Analysis of Numerical Methods, John Wiley, 1966, 541 p.

14. P.J. Laurent (72), Approximation et optimisation, Paris 1972, 531 p.

15. A.R. Mitchell (69), Computational Methods in Partial Differential Equations, John Wiley 1969, 255 p.

16. A.R. Mitchell (73), An Introduction to the Mathematics of the Finite Element Method, in Whiteman (73), 37-58.

17. A.R. Mitchell-R.J.Y. McLeod (73), Curved boundaries in the finite element method, to appear Proc. Symp. Numerical solution of differential equations, Dundee 1973.

18. J.M. Ortega-W.C. Rheinboldt (70), Iterative Solution of nonlinear equations in several variables, Acad. Press (70), 572 p.

19. J. Schrøder (59), Fehlerabschätzungen bei linearen Gleichungssystemen mit dem Brouwerschen Fixpunktsatz, Arch. Rat. Mech. Anal. 3(1959), 28-44.

20. G. Strang (73), The dimension of piecewise polynomial spaces, to appear Proc. Symp. Numerical solution of differential equations, Dundee, 1973.

21. R.S. Varga (62), Matrix iterative analysis, Prentice Hall 1962, 322 p.

22. E.L. Wachspress (73), Algebraic geometry foundations for finite element computation, to appear Proc. Symp. Numerical solution of differential equations, Dundee 1973.

23. W. Walter (70), Differential and Integral Inequalities, Springer 1970, 352 p.

24. J. Werner (70), Einschließungssatz für periodische Lösungen der Liénardschen Differentialgleichung, Computing 5(1970), 246-252.

25. J.R. Whiteman (73), The Mathematics of finite Elements and Applications, Acad. Press, 1973, 520 p.

26. H.J. Wirz (72), Eine Erweiterung des Verfahrens der Zwischenschritte auf allgemeinere parabolische und elliptische Differentialgleichungen, Z. angew. Math. Mech. 52 (1972), 329-336.

27. D.M. Young (73), On the accelerated SSOR method for solving elliptic boundary value problems, to appear Proc. Symp. Numerical solution of differential equations, Dundee 1973.

28. O.C. Zienkiewicz (71), The Finite Element Method in Engineering Science, McGraw Hill, 1971, 521 p.

29. O.C. Zienkiewicz (73), Finite Elements, the Background Story, in Whiteman (73), 1-35.

VARIATIONAL PRINCIPLES - A SURVEY

A.R. Mitchell

University of Dundee, U.K.

A physical problem may be formulated as a variational principle rather than as a differential equation with associated conditions. The basic problem of the variational principle is to determine the function from an admissible class of functions such that a certain definite integral involving the function and some of its derivatives takes on a maximum or minimum value in a closed region R. This is a generalisation of the elementary theory of maxima and minima of the calculus which is concerned with the problem of finding a point in a closed region at which a function has a maximum or minimum value compared with neighboring points in the region. The definite integral in the variational principle is referred to as a functional, since it depends on the entire course of a function rather than on a number of variables. The domain of the functional is the space of the admissible functions. The main difficulty with the variational principle approach is that problems which can be meaningfully formulated as variational principles may not have solutions. This is reflected in mathematical terms by the domain of admissible functions of the functional not forming a closed set. Thus the existence of an extremum (maximum or minimum) cannot be assumed for a variational principle. However, in this text we are concerned with approximate solutions of variational principles. These are obtained by considering some closed subset of the space of permissible functions to provide an upper and lower bound for the theoretical solution of the variational principle.

One apparent advantage of the variational approach is that we seem to require less continuity in the solution function. For example, although the solution of Laplace's equation requires continuity of the derivatives up to the second order in R (C^2

continuity), the solution of the associated variational principle only appears to require the first derivatives to be piecewise continuous in R. As both solutions solve the same problem, it is difficult to see why the variational approach requires less continuity in the solution function. This apparent paradox is explained at length on pages 199-204 of Courant and Hilbert [3] and chapter 2 of Clegg [2]. As a consequence of the weaker continuity requirements, a useful advantage of the variational approach is the greater ease with which approximate solutions can be constructed. A large part of the present text is devoted to describing such approximate methods.

The space over which the integral is evaluated in a variational principle may contain the time co-ordinate. We shall look first at variational principles which do not involve the time. These variational principles, usually of minimum potential energy, govern problems of stable equilibrium which arise from classical field problems of mathematical physics. This chapter contains only the material on variational principles which is relevant to the main theme of this book. No proofs or detailed discussions are given and the interested reader is referred to appropriate parts of books such as Courant and Hilbert [3], Morse and Feshbach [5], Hildebrand [4], Schechter [8], and Clegg [2].

1. STABLE EQUILIBRIUM PROBLEMS

The differential equation which is associated with a variational principle is known as the Euler-Lagrange equation. It is a necessary, but rarely sufficient, condition which a function must satisfy if it is to maximise or minimise a definite integral. The simplest problem of the variational calculus is to determine the minimum of the integral

$$I(u) = \int_{x_0}^{x_1} F(x, u(x), u'(x))dx$$

where the values $u(x_0)$ and $u(x_1)$ are given, and a dash denotes differentiation with respect to x. The necessary but not sufficient condition for the minimum to exist is that u(x) satisfies the differential equation

$$\frac{\partial F}{\partial u} - \frac{d}{dx}\frac{\partial F}{\partial u'} = 0.$$

We shall now generalise this result to cover the following cases;

(1) Two dependent variables. The integral to be minimised is

$$I8u,v) = \int_{x_0}^{x_1} F(x,\ u(x),\ v(x),\ u'(x),\ v'(x))dx$$

where the values $u(x_0)$, $u(x_1)$, $v(x_0)$, $v(x_1)$ are given. The necessary conditions are

$$\frac{\partial F}{\partial u} - \frac{d}{dx}\frac{\partial F}{\partial u'} = 0$$

$$\frac{\partial F}{\partial v} - \frac{d}{dx}\frac{\partial F}{\partial v'} = 0.$$

(2) Two independent variables. The integral is

$$I(u) = \iint_R F(x,y,\ u(x,y),\ u_x(x,y),\ u_y(x,y))dx\ dy$$

where u takes on prescribed values on the boundary of R, the region of integration. The necessary condition is

$$\frac{\partial F}{\partial u} - \frac{\partial}{\partial x}\frac{\partial F}{\partial u_x} - \frac{\partial}{\partial y}\frac{\partial F}{\partial u_y} = 0$$

(3) Higher derivatives. For variational principles involving second derivatives, the integral is

$$I(u,v) = \int_{x_0}^{x_1} F(x,\ u(x),\ v(x),\ u'(x),\ v'(x))dx$$

where the values $u(x_0)$, $u'(x_0)$, $u(x_1)$, $u'(x_1)$ are given and the corresponding necessary condition is

$$\frac{\partial F}{\partial u} - \frac{d}{dx}\frac{\partial F}{\partial u'} + \frac{d^2}{dx^2}\frac{\partial F}{\partial u''} = 0.$$

(4) Constrained extremals. Here the variational problem is constrained by one or more auxiliary conditions. One integral expression is to be made an extremum (maximum or minimum) while one or more other integral expressions maintain fixed values. Such problems are termed <u>isoperimetric</u> problems. Consider for example the problem of determining the function u(x) which maximises (or minimises) the integral

$$I = \int_{x_0}^{x_1} F(x, u(x), u'(x))dx$$

subject to the condition that u(x) satisfies the equation

$$\int_{x_0}^{x_1} G(x, u(x), u'(x))dx = \alpha$$

where the constant α is given. The necessary condition for an extremal to exist is that

$$\frac{\partial(F + \lambda G)}{\partial u} - \frac{d}{dx}\frac{\partial(F + \lambda G)}{\partial u'} = 0, \qquad (1)$$

where the numerical value of the parameter λ is chosen so that (1) is satisfied. A sample example of an isoperimetric problem is the catenary. The problem is to find the shape of a uniform heavy string with fixed end points which hangs under gravity. Here we require to find the function u(x) which passes through the points (x_0, u_0) and (x_1, u_1) and makes the integral

$$\int_{x_0}^{x_1} u(1+u'^2)^{\frac{1}{2}}dx$$

as small as possible, while maintaining a fixed value for the integral

$$\int_{x_0}^{x_1} (1+u'^2)^{\frac{1}{2}}dx.$$

Exercise 1. Show that the length of the curve connecting two points (x_0, u_0) and (x_1, u_1) is

$$I = \int_{x_0}^{x_1} (1+u_x^2)^{\frac{1}{2}}dx.$$

Use the associated Euler-Lagrange equation to find the shortest path between the two points.

Excercise 2. Find the function u(x) which passes through the points (x_0, u_0) and (x_1, u_1) and gives the minimum surface of revolution when rotated about the x-axis.

Exercise 3. Show that

$$\frac{\partial^2 u}{\partial x^2} + \frac{\partial^2 u}{\partial y^2} + \frac{\partial^2 u}{\partial z^2} + f(x,y,z) = 0$$

is the necessary condition for the integral

$$I = \int_V \tfrac{1}{2}[u_x^{\,2} + u_y^{\,2} + u_z^{\,2} - 2u\, f(x,y,z)]dV$$

to be a minimum when u is specified on the surface S which surrounds the volume V.

Before proceeding further, we give some examples of variational principles and equivalent Euler-Lagrange equations. In these examples the region under consideration is R with ∂R as its boundary.

(1) Dirichlet problem for Laplace's equation.

$$I(u) = \int_R \tfrac{1}{2}(u_x^{\,2} + u_y^{\,2})dR \qquad \text{u given on } \partial R.$$

$$u_{xx} + u_{yy} = 0.$$

(2) Loaded and clamped plate. (Biharmonic operator)

$$I(u,v) = \int_R \tfrac{1}{2}[(1-\nu)(u_x^{\,2}+v_y^{\,2}) + \nu(u_x+v_y)^2 + \tfrac{1}{2}(1-\nu)(u_y+v_x)^2]dR$$

$$u_{xxxx} + 2u_{xxyy} + u_{yyyy} = q(x,y)$$

q(x,y) is normal load on the plate.

(3) Small displacement theory of elasticity.

$$I(u,v) = \int_R \tfrac{1}{2}[1-\nu)(u_x^{\,2}+v_y^{\,2}) + \nu(u_x+v_y)^2 + \tfrac{1}{2}(1-\nu)(u_y+v_x)^2]dR$$

u,v given on ∂R.

$$u_{xx} + \nu v_{xy} + \tfrac{1}{2}(1-\nu)(u_{yy}+v_{xy}) = 0$$

$$v_{yy} + \nu u_{xy} + \tfrac{1}{2}(1-\nu)(u_{xx}+v_{xy}) = 0.$$

(4) Radiation (e^u) and Molecular Diffusion (u^2).

$$I(u) = \int_R \tfrac{1}{2}(u_x{}^2+u_y{}^2 + \begin{cases} 2e^u \\ \tfrac{2}{3}u^3 \end{cases})dR \qquad u \text{ given on } \partial R.$$

$$u_{xx} + u_{yy} = \begin{cases} e^u \\ u^2 \end{cases}.$$

(5) Plateau's problem (To find the surface of minimum area bounded by a closed curve in three dimensional space).

$$I(u) = \int_R \tfrac{1}{2}(1 + u_x{}^2 + u_y{}^2)^{\frac{1}{2}}dR \qquad u \text{ given on } \partial R.$$

$$\nabla(\gamma_1)\nabla u = 0 \qquad \gamma_1 = (1 + u_x{}^2 + u_y{}^2)^{-\frac{1}{2}}$$

(6) Non Newtonian Fluids.

$$I(u) = \int_R [\tfrac{1}{2}(u_x{}^2+u_y{}^2)^{1+S} + uc]dR \qquad u \text{ given on } \partial R.$$

$$\nabla(\gamma_2)\nabla u = c \text{ (constant)}, \ \gamma_2 = (u_x{}^2+u_y{}^2)^S \ (0 \geq S \geq -\tfrac{1}{2}).$$

(7) Compressible Flow.

$$I(p) = \int_R p \, dR \qquad \rho \text{ density}$$

p pressure

$$\nabla(\rho \nabla \Phi) = 0. \qquad \Phi \text{ velocity potential}$$

Of the problems above, the first three are linear, the fourth is mildly non-linear, and the last three are non-linear.

2. BOUNDARY CONDITIONS

The steady problems discussed in the previous section have been such that the function is specified on the boundary and so is not subject to variation there. In many problems, however, the function is not specified on the boundary and alternative equally valid boundary conditions apply. Consider, for example, [see Courant and Hilbert [3] pp. 208, 209] the variational problem consisting of the minimisation of the integral

$$I(u) = \int_{x_0}^{x_1} F(x,u,u_x)dx \qquad (2)$$

where u is not specified at the boundary points $x = x_0, x_1$. The necessary conditions for u to minimise I(u) are that u satisfies the Euler Lagrange equation

$$\frac{\partial F}{\partial u} - \frac{d}{dx}\frac{\partial F}{\partial u_x} = 0$$

together with the boundary conditions

$$\frac{\partial F}{\partial u_x} = 0 \qquad \text{at } x = x_0, x_1.$$

The latter are known as natural boundary conditions because they follow directly from the minimisation of the basic integral. If the boundary conditions of the problem consist of neither u specified at the boundary points nor the natural boundary conditions, then the functional to be minimised must be modified in an appropriate manner.

Consider

$$I = \int_{x_0}^{x_1} F(x,u,u_x)dx + [g_1(x,u)]_{x=x_1} - [g_0(x,u)]_{x=x_0}$$

to be the modified form of (2), where $g_0(x,u)$ and $g_1(x,u)$ are unspecified functions. The necessary conditions for this modified integral to have a minimum are [Schecter [8] p. 28]

$$\frac{\partial F}{\partial u} - \frac{d}{dx}\frac{\partial F}{\partial u_x} = 0,$$

together with the boundary conditions

$$\left\{\frac{\partial F}{\partial u_x} + \frac{\partial g_0}{\partial u}\right\}_{x=x_0} = 0,$$

and

$$\left\{ \frac{\partial F}{\partial u_x} + \frac{\partial g_1}{\partial u} \right\}_{x=x_1} = 0,$$

and so the functions $g_0(x,u)$ and $g_1(x,u)$ can be obtained to suit the boundary conditions of the problem. For example the variational principle equivalent to the differential problem consisting of the equation

$$u_{xx} + \Phi(x) = 0$$

together with the boundary conditions

$$-u_x + \alpha u = 0 \qquad \text{at } x = x_0$$

$$u_x + \beta u = 0 \qquad \text{at } x = x_1,$$

is based on the functional

$$I = \int_{x_0}^{x_1} [\tfrac{1}{2}u_x^{\,2} - \Phi(x)u]dx + [\tfrac{1}{2}\beta\, u^2]_{x=x_1} - [\tfrac{1}{2}\alpha\, u^2]_{x=x_0}.$$

If we now consider a variational problem in two space dimensions consisting of the minimisation of the integral

$$I = \int_R F(x,y,u,u_x,u_y)dxdy$$

where u is not specified on the boundary of the region R, the necessary conditions for u to minimise I are that u satisfies the differential equation

$$\frac{\partial F}{\partial u} - \frac{\partial}{\partial x}\frac{\partial F}{\partial u_x} - \frac{\partial}{\partial y}\frac{\partial F}{\partial u_y} = 0$$

together with the natural boundary condition

$$\frac{\partial F}{\partial u_x}\frac{dy}{ds} - \frac{\partial F1}{\partial u_y}\frac{dx}{ds} = 0$$

on the curve ∂R which encloses the region R. If the normal to ∂R makes an angle α with the x-axis, then $\cos\alpha = \frac{dy}{ds}$, and $\sin\alpha = -\frac{dx}{ds}$.

$$[F_{u_x} - \frac{\partial}{\partial x}(F_{u_{xx}})]\frac{dy}{ds} - [F_{u_y} - \frac{\partial}{\partial y}(F_{u_{yy}})]\frac{dx}{ds} - \frac{\partial}{\partial s}\left\{(F_{u_{xx}} - F_{u_{yy}})\frac{dx}{ds}\frac{dy}{ds}\right\}$$

$$+ \frac{1}{2}\frac{\partial}{\partial s} Fu_{xy}\left\{(\frac{dx}{ds})^2 - (\frac{dy}{ds})^2\right\} + \frac{1}{2}\left\{\frac{\partial}{\partial x}(F_{u_{xy}})\frac{dx}{ds} \frac{\partial}{\partial y}(F_{u_{xy}}) \frac{dy}{ds}\right\}$$

$$+ G_u - \frac{\partial}{\partial s}(G_{u_s}) + \frac{\partial^2}{\partial s^2}(G_{u_{ss}}) = 0, \tag{3}$$

and

$$\frac{\partial G}{\partial u_n} + (\frac{dy}{ds})^2 \frac{\partial F}{\partial u_{yy}} + \frac{dx}{ds}\frac{dy}{ds}\frac{\partial F}{\partial u_{xy}} = 0. \tag{4}$$

The function G is chosen so that the boundary conditions given by (3) and (4) correspond to the natural boundary conditions of the problem.

3. TIME DEPENDENT VARIATIONAL PRINCIPLES

The most basic and important time dependent variational principle is Hamilton's principle from which can be deduced the fundamental equations of a large number of physical phenomena. Hamilton's principle states that the motion of a system from time t_0 to time t_1 is such that the time integral of the difference between the kinetic and potential energies is stationary for the true path. This can be expressed in mathematical terms by defining the integral in terms of the Lagrangian L as

$$I = \int_{t_0}^{t_1} Ldt = \int_{t_0}^{t_1} (T-V)dt,$$

where T, V are the respective kinetic and potential energies of the system, and stating that I is made stationary by the actual motion compared with neighboring virtual motions. For a system with n generalised co-ordinates, $q_1, q_2, \text{---} q_n$, the associated Euler-Lagrange equations are

$$\frac{d}{dt}(\frac{\partial T}{\partial \dot{q}_r}) - \frac{\partial}{\partial q_r}(T-V) = 0. \qquad (r = 1,2,\text{---},n)$$

These are usually referred to as Lagrange's equations of motion for the system.

As a simple example of the foregoing theory applied to a continuum, we consider the case of a flexible string under constant tension τ. The string which is fixed at the ends executes small vibrations about the position of stable equilibrium, which is the interval $0 \leq x \leq 1$ of the x-axis. If $u(x,t)$ is the displacement perpendicular to the x-axis of a point on the string, then

$T = \frac{1}{2}\rho \int_0^1 (\frac{\partial u}{\partial t})^2 dx$, and $V = \frac{1}{2}\rho \int_0^1 c^2(\frac{\partial u}{\partial x})^2 dx$, where ρ is the density of the string and $c^2 = \tau/\rho$. The Euler Lagrange equation is

$$\frac{\partial^2 u}{\partial t^2} = c^2 \frac{\partial^2 u}{\partial x^2}$$

which is the wave equation of the string. Thus the wave equation for the string is equivalent to the requirement that the difference between the total kinetic and potential energies of the string be as small as possible, on average, subject to the initial and boundary conditions of the problem.

Other examples of the theory of this section are the vibrating rod, membrane, and plate. (see Courant and Hilbert [3] pp. 244-252.)

We now turn our attention to time dependent dissipative systems, and show how variational principles can be constructed for such systems. The method adopted is to introduce an adjoint system with negative friction. The energy lost by the dissipative system is gained by the adjoint system and so the total energy of the two systems is conserved. For the alternative approach using restricted variational principles, the reader is referred to Rosen [7]. As an example consider the one-dimensional oscillator with friction. Its equation of motion is

$$\ddot{x} + k\dot{x} + n^2 x = 0 \qquad (k < 0)$$

It is required to find a variational principle which has this equation as its Euler Lagrange equation. This is impossible, but if we introduce the adjoint oscillator (represented by the coordinate x^*) with negative friction, its equation of motion is

$$\ddot{x}^* - k\dot{x}^* + n^2 x^* = 0.$$

The purely formal Lagrangian

$$L = \dot{x}\dot{x}^* - \tfrac{1}{2}k(x^*\dot{x} - x\dot{x}^*) - n^2 x\, x^*$$

will be seen to give the above two equations of motion as its Euler Lagrange equations.

Another important example of a dissipative system is the heat diffusion problem. The governing equation for such a problem in one dimension is

$$\frac{\partial u}{\partial t} = \frac{\partial^2 u}{\partial x^2} ,$$

and we introduce the adjoint problem which is governed by the equation

$$- \frac{\partial u^*}{\partial t} \quad \frac{\partial^2 u^*}{\partial x^2} .$$

The formal Lagrangian in this case is

$$L = - \frac{\partial u}{\partial x}\frac{\partial u^*}{\partial x} - \frac{1}{2} u^* \frac{\partial u}{\partial t} - u \frac{\partial u^*}{\partial t} ,$$

which gives the above two equations as its Euler Lagrange equations.

4. DUAL VARIATIONAL PRINCIPLES

So far our variational principles have been one sided i.e. the approximation solution always lies above or below the theoretical solution of the variational principle. It is often possible, however, to construct two variational principles for a problem, where the same quantity d is a minimum and maximum with respect to the two principles. If d^u and d^l are approximate solutions of the minimum and maximum principles respectively, then

$$d^l \leq d \leq d^u,$$

and so we have a practical method of bounding d. It is to be hoped that the quantity d is of physical significance.

Some examples will now be given of problems for which dual variational principles can be constructed.

(1) The classical Dirichlet problem. Here the functional

$$I(u) = \int_R \frac{1}{2} \left\{ u_x^2 + u_y^2 \right\} \; dR$$

is minimised with respect to continuous functions $u(x,y)$ which have piecewise continuous derivatives in the region R and take prescribed values $u = f(s)$ on ∂R, where s is the arc length of ∂R, the boundary of R. The complementary or dual problem is the functional

$$J(v) = -\int_R \tfrac{1}{2}\{v_x^2 + v_y^2\}\, dR - \int_{\partial R} vf'(s)ds$$

maximised with respect to continuous functions $v(x,y)$ which have piecewise continuous derivatives in R and satisfy natural boundary conditions on ∂R. In this example

$$\min_u I(u) = \max_v J(v) = d.$$

Exercise 4. An incompressible inviscid flow is parallel to the x-axis. Calculate $I(u)$ and $J(v)$ for this problem where R is the square region $0 \le x,y \le 1$, and show that the extreme values coincide. [Hint. The functions u and v are the stream function and potential respectively for the flow.]

Exercise 5. Show that the necessary conditions for $J(v)$ to have a maximum consist of the Euler Lagrange equation

$$v_{xx} + v_{yy} = 0,$$

together with the natural boundary condition

$$v_y \frac{dx}{ds} - v_x \frac{dy}{ds} = f'(s).$$

(2) Small displacement theory of elasticity. (Washizu [11]). Consider an isotropic body in three dimensional space occupying a region R enclosed by a surface ∂R. The components of the body forces per unit volume are (X,Y,Z). The surface of the body is divided into two parts, S_σ where the boundary conditions consist of external forces $(\bar{X},\bar{Y},\bar{Z})$ per unit area, and S_u over which the displacements $(\bar{u},\bar{v},\bar{w})$ are given. We have $\partial R = S_\sigma + S_u$. Now the total potential energy is given by

$$\Pi = \int_R W(\epsilon_x,\epsilon_y,\epsilon_z,\gamma_{yz},\gamma_{xy})dR - \int_R (Xu+Yv+Zw)dR - \int_{S_\sigma} (\bar{X}u+\bar{Y}v+\bar{Z}w)d(\partial R),$$

$$W = \frac{E\nu}{2(1+\nu)(1-2\nu)} (u_x + v_y + w_z)^2 + \frac{E}{2(1+\nu)} (u_x{}^2 + v_y{}^2 + w_z{}^2)$$

$$+ \frac{E}{4(1+\nu)} [(v_z + w_y)^2 + (w_x + u_z)^2 + (u_y + v_x)^2].$$

If the body forces and the surface forces are kept unchanged during variation, Π is a minimum due to the actual displacements. This is the principle of minimum potential energy.

The complementary energy is given by

$$\Pi_c = \int_R \Phi(\sigma_x,\sigma_y,\sigma_z,\tau_{yz},\tau_{zx},\tau_{xy})dR - \int_{S_u} (X\bar{u} + Y\bar{v} + Z\bar{w})d(\partial R),$$

where

$$\Phi = \frac{1}{2E} [(\sigma_x + \sigma_y + \sigma_z)^2 + 2(1+\nu)(\tau_{yz}{}^2 + \tau_{xy}{}^2 - \sigma_y\sigma_z - \sigma_z\sigma_x - \sigma_x\sigma_y)].$$

If the surface displacements are kept unchanged during variation, Π_c is a minimum due to the actual stresses. This is the principal of minimum complementary energy.

The quantity which can be bounded conveniently by these two principles is the direct influence coefficient or generalised displacement. (See Pian [6]).

Exercise 6. Show that $W = \Phi$ if the following linear stress strain relations hold

$$E\epsilon_x = \sigma_x - \nu(\sigma_y + \sigma_z) \qquad \tau_{yz} = \frac{E}{2(1+\nu)} \gamma_{yz}$$

$$E\epsilon_y = \sigma_y - \nu(\sigma_x + \sigma_z) \qquad \tau_{zx} = \frac{E}{2(1+\nu)} \gamma_{zx}$$

$$E\epsilon_z = \sigma_z - \nu(\sigma_x + \sigma_y) \qquad \tau_{xy} = \frac{E}{2(1+\nu)} \gamma_{xy}.$$

Exercise 7. Show that the necessary conditions for the potential energy

$$\Pi = \int_R W(u,v)dR$$

where

$$W(u,v) = \frac{E\nu}{2(1+\nu)(1-2\nu)}(u_x+v_y)^2 + \frac{E}{2(1+\nu)}(u_x{}^2+v_y{}^2) + \frac{E}{4(1+\nu)}(u_y+v_x)^2$$

to have a minimum are the Euler Lagrange equations

$$(2-2\nu)u_{xx} + (1-2\nu)u_{yy} + v_{xy} = 0$$

$$(2-2\nu)v_{yy} + (1-2\nu)v_{xx} + u_{xy} = 0$$

in the region R together with the boundary conditions

$$(2-2\nu)u_x \cos\alpha+(1-2\nu)u_y \sin\alpha+(1-2\nu)v_x \sin\alpha + 2\nu v_y \cos\alpha = 0$$

$$2\nu u_x \sin\alpha+(1-2\nu)u_y \cos\alpha+(1-2\nu)v_x \cos\alpha+(2-2\nu)v_y \sin\alpha = 0$$

on ∂R.

(3) Compressible flow. (Sewell [10]). The respective volume integrands which appear in the dual variational principles are the pressure p, and the quantity $p + \rho v^2$, where ρ is the density and v is the velocity of the fluid. Here

$$p = p(v_i, h, \eta)$$

where $v_i (i = 1,2,3)$ are the velocity components, h and η are the total energy and entropy per unit mass respectively. The results

$$\frac{\partial p}{\partial v_i} = - Q_i (i = 1,2,3), \quad \frac{\partial p}{\partial h} = \rho, \quad \frac{\partial p}{\partial \eta} = - \rho T$$

follow where $Q_i = \rho v_i$ $(i = 1,2,3)$ with ρ the density and T the temperature. The function

$$p = P(Q_i, h, \eta) = \sum_{i=1}^{3} Q_i v_i + p$$

is introduced, where

$$\frac{\partial P}{\partial Q_i} = v_i (i = 1,2,3), \quad \frac{\partial P}{\partial h} = \rho, \quad \frac{\partial P}{\partial \eta} = - \rho T.$$

The dual variational principles involving p and P respectively can be strengthened to extremum principles for particular types of compressible flow.

Two recent accounts attempting to unify dual principles are due to Sewell [10] and Arthurs [1]. The former uses Legendre or involutory transformations, and the latter uses the Canonial Theory of the Euler Lagrange equation. A brief account of the Legendre transformation is now given, and the reader who wishes to take this interesting topic further is referred to the paper by Sewell.

Legendre transformations. The transition from one variational principle to another is often a natural consequence of a Legendre dual transformation. Consider the function

$$Y = Y(y_i, u_\alpha)$$

of n active variables $y_i (i = 1,2,---,n)$ and m passive variables u_α $(\alpha = 1,2,---,m)$. Another set of variables $x_i (i = 1,2,---,n)$ is introduced by the transformation

$$x_i = \frac{\partial Y}{\partial y_i} \quad (i = 1,2,---,n)$$

If this transformation is single valued and reversible, the inverse transformation may be written as

$$y_i = \frac{\partial X}{\partial x_i} \quad (i = 1,2,---,n)$$

for a scalar function $X = X(x_i, u_\alpha)$ defined by

$$X = \sum_{i=1}^{n} x_i y_i - Y.$$

The passive variables have the property

$$\frac{\partial X}{\partial u_\alpha} = - \frac{\partial Y}{\partial u_\alpha} \quad (\alpha = 1,2,---,m).$$

Three particular examples of this general theory are

(i) Small displacement theory of elasticity.

$$n = 6 \quad x_i \sim \sigma_x, \sigma_y, \sigma_z, \tau_{yz}, \tau_{zx}, \tau_{xy}$$

$$y_i \sim \epsilon_x, \epsilon_y, \epsilon_z, \gamma_{yz}, \gamma_{zx}, \gamma_{xy}$$

$m = 3 \qquad u_\alpha \sim u, v, w$

$$X \equiv \Phi + (Xu + Yv + Zw)$$

$$Y \equiv W - (Xu + Yv + Zw)$$

(ii) Compressible flow.

$n = 3 \qquad x_i \sim \rho v_1, \rho v_2, \rho v_3$

$y_i \sim v_1, v_2, v_3$

v_i velocity components, ρ density.

$n = 2 \qquad u_\alpha \sim h, \eta$

h total energy per unit mass, η entropy per unit mass.

$X \equiv p + \rho v^2 \qquad v \equiv (v_1, v_2, v_3)$

$Y \equiv -p \qquad$ p pressure.

(iii) Classical dynamics.

$n = 3 \qquad x_i \sim p_1, p_2, p_3$ momenta

$y_i \sim \dot{q}_1, \dot{q}_2, \dot{q}_3$ velocities

$m = 4 \qquad u_\alpha \sim q_1, q_2, q_3, t$

$X \equiv H$ Hamiltonian

$Y \equiv L$ Lagrangian.

Finally, an excellent recent account of dual variational principles can be found in Noble and Sewell.

Remark. The search for new variational principles is tremendously important. In continuum mechanics, the extension of Hamilton's principle is easy in the Lagrangian description. This is not so in the Eulerian case. An interesting account on the construction of variational principles for Eulerian formulations of problems is given by Seliger and Whitham [9].

References

1. Arthurs, A M Complementary Variational Principles
Oxford Clarendon Press 1970.

2. Clegg, J C Calculus of Variations
Oliver and Boyd 1967.

3. Courant, R and Hilbert, D Methods of Mathematical Physics
Interscience 1953.

4. Hildebrand, F B Methods of Applied Mathematics
Prentice Hall 1965.

5. Morse, P M and Feshbach, H Methods of Theoretical Physics
McGraw Hill 1953.

6. Pian, T H H Numerical Solution of Field Problems in
Continuum Physics
SIAM - AMS Proceedings Vol. 2 1970.

7. Rosen P J. Chem. Phys. 21, 1220. 1953
J. App. Phys. 25, 336. 1954.

8. Schechter, R S The Variational Method in Engineering
McGraw Hill 1967.

9. Seliger, R L and Whitham, G B Proc. Roy. Soc. (London)
A 305, 1. 1968.

10. Sewell, M J Phil. Trans. Roy. Soc. (London) A 265, 319 1969.

11. Washizu, K Variational Methods in Elasticity and Plasticity
Pergamon Press 1968.

SOLUTION OF LINEAR SYSTEMS OF EQUATIONS

David M. Young*

The University of Texas at Austin, U.S.A.

1. INTRODUCTION

In this paper we consider various iterative methods for solving systems of linear algebraic equations. We shall be primarily concerned with large systems with sparse matrices such as arise in the solution of elliptic boundary value problems in two dimensions by finite difference methods. Our discussion is divided into two parts: First, we review the well-known facts about such methods as the Jacobi, Gauss-Seidel, and successive overrelaxation methods. A treatment of various acceleration techniques is included. In the second part of the paper we consider the symmetric overrelaxation method and describe some practical procedures which can be used to obtain very rapid convergence. By showing that the method can be very effective in many cases and by outlining a definite procedure for its application, we hope to encourage its wider usage and to stimulate further research.

2. A MODEL PROBLEM

To facilitate the discussion let us consider linear systems arising from the finite difference solution of the following model problem. Given a function g(x,y) defined and continuous on the boundary S of the unit square, we seek a function u(x,y) which is twice differentiable in the interior R of the square and continu-

---

*Work on this paper was supported in part by the U.S.Army Research Office (Durham), grant DA-ARO-D-31-124-734 at The University of Texas at Austin.

ous in R+S such that

$$(2.1)\qquad \frac{\partial^2 u}{\partial x^2} + \frac{\partial^2 u}{\partial y^2} = 0$$

in R and

$$(2.2)\qquad u(x,y) = g(x,y)$$

on S. We consider the method of finite differences where for each point of R_h we require that u(x,y) satisfy the difference equation

$$(2.3)\qquad \frac{u(x+h,y)+u(x,y+h)+u(x-h,y)+u(x,y-h)-4u(x,y)}{h^2} = 0.$$

We also require that (2.2) holds for points of S_h. Here for each $h > 0$ such that h^{-1} is an integer we let R_h and S_h be those points (ph, qh), with p and q integers, belonging to R and S, respectively.

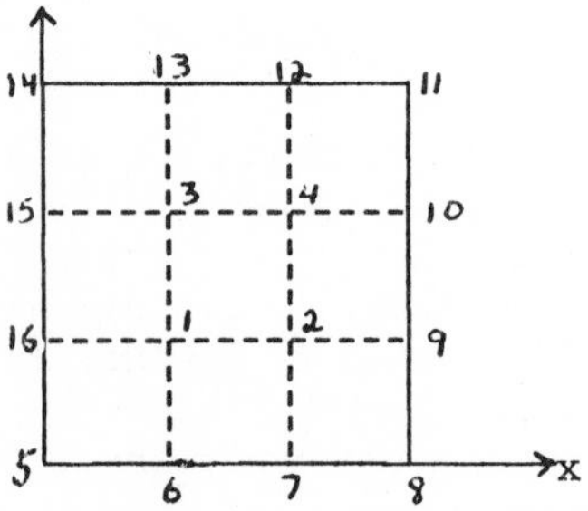

Figure 2.1 A Model Problem

As an example, consider the case h = 1/3. Numbering the mesh points as in Figure 2.1 and multiplying the difference equation by $-h^2$ we get

$$(2.4)\qquad \begin{pmatrix} 4 & -1 & -1 & 0 \\ -1 & 4 & 0 & -1 \\ -1 & 0 & 4 & -1 \\ 0 & -1 & -1 & 4 \end{pmatrix} \begin{pmatrix} u_1 \\ u_2 \\ u_3 \\ u_4 \end{pmatrix} = \begin{pmatrix} g_6 + g_{16} \\ g_7 + g_9 \\ g_{13} + g_{15} \\ g_{12} + g_{10} \end{pmatrix} = \begin{pmatrix} b_1 \\ b_2 \\ b_3 \\ b_4 \end{pmatrix}$$

For large values of h^{-1} we get a large number of equations. However, the matrix is very sparse since there are never more than five nonzero elements in any one row. Because of this sparseness and because of the large number of equations, iterative methods have often been used in preference to direct methods such as the Gauss elimination method. However, we should remark that the size of problem for which direct methods are being used has increased steadily over the years, especially for systems having certain special properties.

3. BASIC ITERATIVE METHODS

We now define three "basic" iterative procedures, namely, the Jacobi, Gauss-Seidel, and successive overrelaxation methods. We first write the system (2.4) in the matrix form

$$Au = b. \tag{3.1}$$

Next we write (3.1) in the equivalent form

$$u = Bu + c \tag{3.2}$$

where

$$B = \begin{pmatrix} 0 & 1/4 & 1/4 & 0 \\ 1/4 & 0 & 0 & 1/4 \\ 1/4 & 0 & 0 & 1/4 \\ 0 & 1/4 & 1/4 & 0 \end{pmatrix} \tag{3.3}$$

and $c = (1/4)b$. In general we let

$$B = I - D^{-1}A \tag{3.4}$$

where D is a diagonal matrix with the same diagonal elements as A.

The Jacobi method is defined as follows. Starting with an arbitrary initial vector $u^{(0)}$ we define $u^{(1)}, u^{(2)}, \ldots$ by

$$u^{(n+1)} = Bu^{(n)} + c. \tag{3.5}$$

Thus, for the system (2.4) we have

$$\begin{cases} u_1^{(n+1)} = \frac{1}{4}u_2^{(n)} + \frac{1}{4}u_3^{(n)} + c_1 \\ u_2^{(n+1)} = \frac{1}{4}u_1^{(n)} + \frac{1}{4}u_4^{(n)} + c_2 \end{cases} \tag{3.6}$$

etc. Note that even though $u_1^{(n+1)}$ is available when we compute $u_2^{(n+1)}$, we use the old value $u_1^{(n)}$ instead. With the Gauss-Seidel method, we use new values as soon as available. Thus we have

$$(3.7)\quad \begin{cases} u_1^{(n+1)} = \frac{1}{4}u_2^{(n)} + \frac{1}{4}u_3^{(n)} + c_1 \\ u_2^{(n+1)} = \frac{1}{4}u_1^{(n+1)} + \frac{1}{4}u_4^{(n)} + c_2 \end{cases}$$

etc. This can be written in matrix form as

$$(3.8)\quad u^{(n+1)} = Lu^{(n+1)} + Uu^{(n)} + c$$

where L and U are strictly lower triangular and strictly upper triangular matrices respectively such that

$$(3.9)\quad L + U = B.$$

We note that even though (3.8) appears to define $u^{(n+1)}$ implicitly, nevertheless, the calculation can be carried out explicitly. For purpose of analysis of the Gauss-Seidel method, we write (3.8) in the form

$$(3.10)\quad u^{(n+1)} = \mathcal{L}u^{(n)} + (I-L)^{-1}c.$$

where

$$(3.11)\quad \mathcal{L} = (I-L)^{-1}U.$$

The rapidity of convergence of a linear stationary iterative method of first degree of the form

$$(3.12)\quad u^{(n+1)} = Gu^{(n)} + k$$

depends on the spectral radius of G, which we denote by S(G). Roughly speaking, the error is reduced by a factor of S(G) by each iteration. We also define the <u>reciprocal rate of convergence</u> of the method by

$$(3.13)\quad RR(G) = \frac{1}{-\log S(G)} .$$

Roughly speaking, the number of iterations required to achieve a certain level of convergence is proportional to RR(G).

For the model problem it is easy to show that

$$(3.14)\quad \begin{cases} S(B) = \cos \pi h \\ S(\mathcal{L}) = \cos^2 \pi h \end{cases}$$

so that for small h

(3.15) $$\begin{cases} RR(B) \doteq \dfrac{2}{\pi^2} h^{-2} \\ RR(\mathcal{L}) \doteq \dfrac{1}{\pi^2} h^{-2}. \end{cases}$$

Thus the number of iterations, $\mathcal{N}$, required for convergence increases as h^{-2}. This rapid increase in $\mathcal{N}$ makes these two methods unsuitable except for very small problems.

The convergence can often be accelerated by modifying the Gauss-Seidel method slightly. We introduce a "relaxation factor," ω, and use the formulas

(3.16) $$\begin{cases} u_1^{(n+1)} = \omega \left\{\frac{1}{4}u_2^{(n)} + \frac{1}{4}u_3^{(n)} + c_1\right\} + (1-\omega)u_1^{(n)} \\ u_2^{(n+1)} = \omega \left\{\frac{1}{4}u_1^{(n+1)} + \frac{1}{4}u_4^{(n)} + c_2\right\} + (1-\omega)u_2^{(n)} \end{cases}$$

etc. This gives the <u>successive overrelaxation</u> method (SOR method). In matrix form we have

(3.17) $$u^{(n+1)} = \omega(Lu^{(n+1)} + Uu^{(n)} + c) + (1-\omega)u^{(n)}$$

or, equivalently,

(3.18) $$u^{(n+1)} = \mathcal{L}_\omega u^{(n)} + (I-\omega L)^{-1}\omega c$$

where

(3.19) $$\mathcal{L}_\omega = (I-\omega L)^{-1}(\omega U + (1-\omega)I).$$

For systems corresponding to the model problem and, more generally, for systems where the matrix A is positive definite and consistently ordered* the optimum value of ω is given by

(3.20) $$\omega_b = \frac{2}{1+\sqrt{1-S(B)^2}} .$$

Using ω_b one obtains an order-of-magnitude improvement in the rate of convergence. In fact, for the model problem we have

(3.21) $$RR(\mathcal{L}_{\omega_b}) \doteq \frac{1}{2\pi} h^{-1}.$$

Thus the number of iterations increases only with the first power of h^{-1}.

*See, for instance, Young [16].

We remark that almost as great an improvement in convergence rate can be obtained using the SOR method with ω given in (3.20) for a linear system whose matrix is a positive definite L-matrix* (i.e., a Stieltjes matrix). This was shown by Kahan [7].

4. CHOICE OF RELAXATION FACTOR

The effective application of the SOR method depends upon the availability of a good choice of ω or, equivalently, by (3.20), of a good estimate of S(B). As shown in [16], a slight underestimate of S(B) and ω results in a much greater loss of convergence speed than does an equivalent overestimate. Young [16,17] gave the following upper bound for S(B),

$$(4.1)\qquad S(B) \le \frac{2(\bar{A}+\bar{C})}{2(\bar{A}+\bar{C})+h^2(\underline{-F})} \times \left\{1 - \frac{2\underline{A}\sin^2\frac{\pi}{2I} + 2\underline{C}\sin^2\frac{\pi}{2J}}{\frac{1}{2}(\bar{A}+\underline{A}) + \frac{1}{2}(\bar{C}+\underline{C}) + \frac{1}{2}(\bar{A}-\underline{A})\cos\frac{\pi}{I} + \frac{1}{2}(\bar{C}-\underline{C})\cos\frac{\pi}{J}}\right\} .$$

Here $\underline{A} \le A(x,y) \le \bar{A}$, $\underline{C} \le C(x,y) \le \bar{C}$, and $-\bar{F}(x,y) \ge -F(x,y) \ge (\underline{-F})$ in R+S. This estimate is applicable to a problem involving the self-adjoint differential equation

$$(4.2)\qquad \frac{\partial}{\partial x}\left(A\frac{\partial u}{\partial x}\right) + \frac{\partial}{\partial y}\left(C\frac{\partial u}{\partial y}\right) + Fu = G$$

where in R+S the conditions

$$(4.3)\qquad A(x,y) > 0,\ C(x,y) > 0,\ F(x,y) \le 0$$

are satisfied. The symmetric difference equation

$$(4.4)\qquad h^{-1}\left\{A(x+\tfrac{h}{2},h)\left[\frac{u(x+h,y)-u(x,y)}{h}\right] - A(x-\tfrac{h}{2},y)\left[\frac{u(x,y)-u(x-h,y)}{h}\right]\right\}$$
$$+ h^{-1}\left\{C(x,y+\tfrac{h}{2})\left[\frac{u(x,y+h)-u(x,y)}{h}\right] - C(x,y-\tfrac{h}{2})\left[\frac{u(x,y)-u(x,y-h)}{h}\right]\right\}$$
$$+ Fu(x,y) = G(x,y)$$

is used. (We assume that the boundary of the region consists of horizontal and vertical lines such that for any mesh point of R the four neighboring points are in R or else lie on S. This is assumed to be true for a sequence of values of h tending to zero.)

---

*As defined in [16], an L-matrix has positive diagonal elements and nonpositive off-diagonal elements.

It is assumed that R+S is circumscribed by an Ih x Jh rectangle. (See Figure 4.1.)

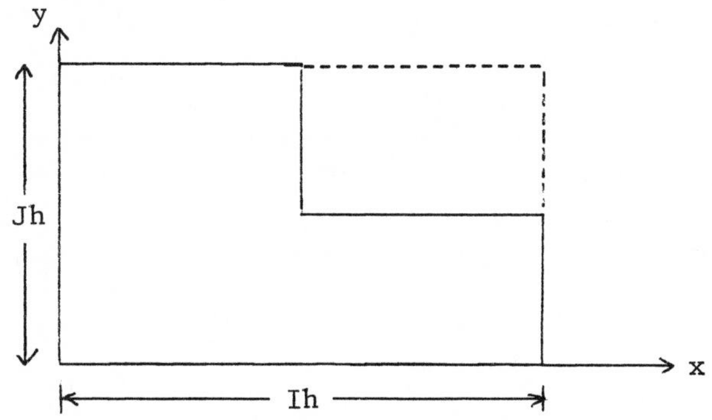

Figure 4.1. A Circumscribing Rectangle

The above procedure for estimating S(B) (and ω_b) is an "a priori" procedure as contrasted to an "a postiori" procedure wherein S(B) is determined dynamically. A very simple "a postiori" procedure was used by Young [14] and involves iterating with the Gauss-Seidel method and estimating $S(\mathcal{L})$, which equals $S(B)^2$, based on the results of the iteration process. More refined procedures have been used by Hageman and Kellogg [6], Carré [2], Kulsrud [8], Rigler [11], Reid [9], and Wachspress [13].

5. CONVERGENCE ACCELERATION

The convergence of a linear stationary iterative method of first degree of the form (3.12) can be accelerated by an order of magnitude provided that the eigenvalues μ of the iteration matrix G are real and lie in the interval

$$(5.1) \qquad \alpha \leq \mu \leq \beta < 1.$$

Such is the case for the Jacobi method when A is positive definite and for the Gauss-Seidel method when A is positive definite and consistently ordered. It is unfortunately not the case, in general, for the SOR method.

The acceleration can be achieved by means of variable second-degree, stationary second-degree, or variable extrapolation methods. The optimum nonstationary second-degree method is equivalent to the optimum "semi-iterative" method (see, for instance, Golub and Varga [5]). The method is defined by

$$(5.2) \qquad u^{(n+1)} = \rho_{n+1}(\bar{\rho}(Gu^{(n)}+k) + (1-\bar{\rho})u^{(n)}) + (1-\rho_{n+1})u^{(n-1)}.$$

Here

$$(5.3) \qquad \bar{\rho} = \frac{2}{2-(\beta+\alpha)}$$

$$(5.4) \qquad \begin{cases} \rho_1 = 1 \\ \rho_2 = \dfrac{1}{1-\frac{\sigma^2}{2}} \\ \rho_{n+1} = \dfrac{1}{1-\frac{\sigma^2}{4}\rho_n}, \qquad n = 2, 3, \ldots \end{cases}$$

$$(5.5) \qquad \sigma = \frac{\beta-\alpha}{2-(\beta+\alpha)} .$$

To measure the effectiveness of the method we write (5.2) in the form

$$(5.6) \qquad u^{(n)} = \mathcal{G}_n u^{(0)} + k_n$$

where $\mathcal{G}_n$ is a suitable polynomial in G. It can be shown that

$$(5.7) \qquad S(\mathcal{G}_n) = \frac{2r^{n/2}}{1+r^n}$$

where

$$(5.8) \qquad r = \left(\frac{\sigma}{1+\sqrt{1-\sigma^2}}\right)^2 .$$

Also, the <u>reciprocal average rate of convergence</u> which is given by

$$(5.9) \qquad RR(\mathcal{G}_n) = \frac{1}{-(1/n)\log S(\mathcal{G}_n)}$$

approaches the <u>reciprocal asymptotic average rate of convergence</u>

$$(5.10) \qquad RR_\infty(\mathcal{G}_n) = \frac{1}{-(1/2)\log r}$$

as $n \to \infty$.

It can be shown that for σ close to unity we have

$$(5.11)\quad RR_\infty(\mathcal{G}_n) = \frac{1}{-(1/2)\log r} \sim \frac{1}{\sqrt{2}}\sqrt{\frac{1}{-\log \sigma}}$$
$$\leq \frac{1}{\sqrt{2}}\sqrt{\frac{1}{-\log S(G)}} = \frac{1}{\sqrt{2}}\sqrt{RR(G)}\ .$$

Thus the reciprocal asymptotic average rate of convergence of the accelerated method is smaller, by an order of magnitude, than the reciprocal rate of convergence of the basic method.

For example, for the Jacobi method and the model problem we have $\alpha = -\cos \pi h$, $\beta = \cos \pi h$, $\sigma = \cos \pi h$, and

$$(5.12)\quad RR_\infty(\mathcal{G}_n) \doteq \frac{1}{\pi} h^{-1}.$$

This is twice $RR(\mathcal{L}_{\omega_b})$. However, by ordering the mesh points in the "red-black" ordering it is possible to double the convergence rate by considering only half of the mesh points on each iteration. This gives the <u>cyclic Chebyshev semi-iterative (CCSI) method</u> (Golub and Varga [5]), which has approximately the same convergence rate as the SOR method. The CCSI method is actually superior to the SOR method when measured in terms of certain matrix norms.

For the Gauss-Seidel method we have $\alpha = 0$, $\beta = S(B)^2$, and

$$(5.13)\quad RR_\infty(\mathcal{G}_n) \doteq \frac{1}{2\pi} h^{-1}.$$

Thus the Gauss-Seidel semi-iterative method converges exactly twice as fast as the Jacobi semi-iterative method. Thus it converges approximately as fast as the SOR method. It is recommended that the "red-black ordering" be used--otherwise with the "natural ordering" numerical instability can occur.

We remark that if we replace $\rho_2, \rho_3, \ldots$ in (5.2) by

$$(5.14)\quad \rho_\infty = \frac{2}{1+\sqrt{1-\sigma^2}}.$$

then we obtain a <u>stationary second-degree</u> method. The convergence of the stationary method is nearly as good as that of the non-stationary method (see Golub [4] and Young [16,18]).

As an alternative to a second-degree method we can obtain nearly as great an acceleration of convergence using variable extrapolation. This has the advantage of saving computer memory since one is not required to have available $u^{(n-1)}$ as well as $u^{(n)}$ in computing $u^{(n+1)}$. The variable extrapolation method is based on Richardson's method [10]. The formulas are

$$(5.15)\quad u^{(n+1)} = \theta_{n+1}(Gu^{(n)} + k) + (1 - \theta_{n+1})u^{(n)}$$

where

$$(5.16)\quad \theta_k = \frac{2}{2-(\beta-\alpha)\cos\frac{(2k-1)\pi}{2m} - (\beta+\alpha)}, \quad k = 1,2,\ldots,m.$$

Here one chooses an integer m and uses the θ_k in the cyclic order $\theta_1,\theta_2,\ldots,\theta_m,\theta_1,\theta_2,\ldots$. The larger one chooses m, the closer the reciprocal average rate of convergence will be to that of the semi-iterative method. However, m should not be taken too large for two reasons: first, we can only expect convergence on the m-th, 2m-th, 3m-th,... iterations; second, numerical instability may occur if m is too large (see Young [15]). This is more serious for the Jacobi and Gauss-Seidel methods than for the symmetric SOR method which we shall consider below.

To summarize, for the model problem we have the following reciprocal asymptotic average rates of convergence.

Method	Matrix	RR(G)
Jacobi	B	$\frac{2}{\pi^2} h^{-2}$
Gauss-Seidel	$\mathcal{L}$	$\frac{1}{\pi^2} h^{-2}$
SOR	$\mathcal{L}_{\omega_b}$	$\frac{1}{2\pi} h^{-1}$
Jacobi semi-iterative		$\frac{1}{\pi} h^{-1}$
CCSI (red-black ordering)		$\frac{1}{2\pi} h^{-1}$
Gauss-Seidel semi-iterative (red-black ordering recommended)		$\frac{1}{2\pi} h^{-1}$

6. THE SYMMETRIC SUCCESSIVE OVERRELAXATION METHOD

In the second part of the paper we shall consider the use of the symmetric successive overrelaxation method (SSOR method). This method was first considered by Sheldon [12]. For a large class of problems the accelerated SSOR method affords a large improvement in convergence over the SOR method. Moreover, a simple test is available to determine in advance whether or not the method will be effective. By showing that the method is potentially effective and that "a priori" predictions are possible, we hope to encourage the wider use of the method.

Each iteration of the SSOR method consists of two half iterations. The first half iteration is the ordinary (forward) SOR method, while the second half iteration is the backward SOR method. Thus we have by (3.17)

$$(6.1)\quad \begin{cases} u^{(n+\frac{1}{2})} = \omega\left[Lu^{(n+\frac{1}{2})} + Uu^{(n)} + c\right] + (1-\omega)u^{(n)} \\ u^{(n+1)} = \omega\left[Lu^{(n+\frac{1}{2})} + Uu^{(n+1)} + c\right] + (1-\omega)u^{(n+\frac{1}{2})}. \end{cases}$$

Let us give specific formulas for the SSOR method for linear systems corresponding to the difference equation (4.4), which we now write in the form

$$(6.2)\quad u(x,y) = \beta_1(x,y)u(x+h,y) + \beta_2(x,y)u(x,y+h) + \beta_3(x,y)u(x-h,y) + \beta_4(x,y)u(x,y-h) + \tau(x,y)$$

where

$$(6.3)\quad \begin{cases} \beta_1(x,y) = \dfrac{A(x+\frac{h}{2},y)}{S(x,y)}, & \beta_2(x,y) = \dfrac{C(x,y+\frac{h}{2})}{S(x,y)} \\ \beta_3(x,y) = \dfrac{A(x-\frac{h}{2},y)}{S(x,y)}, & \beta_4(x,y) = \dfrac{C(x,y-\frac{h}{2})}{S(x,y)} \end{cases}$$

and where

$$(6.4)\quad \begin{cases} S(x,y) = A(x+\frac{h}{2},y) + A(x-\frac{h}{2},y) + C(x,y+\frac{h}{2}) + C(x,y-\frac{h}{2}) - h^2F(x,y) \\ \tau(x,y) = -h^2\dfrac{G(x,y)}{S(x,y)}. \end{cases}$$

The formulas for the SSOR method can be written in the form

$$(6.5)\quad \begin{cases} u^{(n+\frac{1}{2})}(x,y) = \omega\Big\{\beta_1 u^{(n)}(x+h,y) + \beta_2 u^{(n)}(x,y+h) \\ \qquad + \beta_3 u^{(n+\frac{1}{2})}(x-h,y) + \beta_4 u^{(n+\frac{1}{2})}(x,y-h) + \tau(x,y)\Big\} \\ \qquad + (1-\omega)u^{(n)}(x,y) \end{cases}$$

(6.5) cont'd

$$\begin{cases} u^{(n+1)}(x,y) = \omega\Big\{\beta_1 u^{(n+1)}(x+h,y) + \beta_2 u^{(n+1)}(x,y+h) \\ \quad + \beta_3 u^{(n+\frac{1}{2})}(x-h,y) + \beta_4 u^{(n+\frac{1}{2})}(x,y-h) + \tau(x,y)\Big\} \\ \quad + (1-\omega)u^{(n+\frac{1}{2})}(x,y) \end{cases}$$

To analyze the convergence of the SSOR method we eliminate $u^{(n+\frac{1}{2})}$ from (6.1) obtaining

$$u^{(n+1)} = \mathcal{S}_\omega u^{(n)} + \omega(2-\omega)(I-\omega U)^{-1}(I-\omega L)^{-1}D^{-1}b \tag{6.6}$$

where

$$\mathcal{S}_\omega = \mathcal{U}_\omega \mathcal{L}_\omega = I - \omega(2-\omega)(I-\omega U)^{-1}(I-\omega L)^{-1}D^{-1}A. \tag{6.7}$$

Here $\mathcal{L}_\omega$ is given by (3.19) and

$$\mathcal{U}_\omega = (I-\omega U)^{-1}(\omega L+(1-\omega)I). \tag{6.8}$$

It can be shown (see, for instance [16]) that the eigenvalues of $\mathcal{S}_\omega$ are real, non-negative, and less than unity provided that A is positive definite and $0 < \omega < 2$. Therefore, the SSOR method is convergent. Moreover, from the analysis of Section 5, it follows that an order-of-magnitude improvement in the convergence rate can be obtained using acceleration. Using results given in [19], we now show that the reciprocal rate of convergence of the SSOR method for the system (6.5) is $O(h^{-1})$. From this and from (5.11), it follows that the reciprocal asymptotic rate of convergence of the accelerated procedure is $O(h^{-\frac{1}{2}})$.

Given bounds $\bar{\mu}$ and $\bar{\beta}$ for S(B) and S(LU), respectively, one can give a bound in $S(\mathcal{S}_\omega)$ for any ω. It is easy to show that $S(B) \leq 2\sqrt{S(LU)}$; hence, if $\bar{\mu} > 2\sqrt{\bar{\beta}}$ one should replace $\bar{\mu}$ by $2\sqrt{\bar{\beta}}$. In [19] it is shown that if A has Property A (see [16]), then

$$S(\mathcal{S}_\omega) \leq \begin{cases} 1-\omega(2-\omega)\dfrac{1-\bar{\mu}}{1-\omega\bar{\mu}+\omega^2\bar{\beta}} & \text{if } \bar{\beta} \geq \frac{1}{4} \text{ or } \bar{\beta} < \frac{1}{4} \text{ and } \omega \leq \omega^* \\ 1-\omega(2-\omega)\dfrac{1+\bar{\mu}}{1+\omega\bar{\mu}+\omega^2\bar{\beta}} & \text{if } \bar{\beta} < \frac{1}{4} \text{ and } \omega > \omega^*. \end{cases} \tag{6.9}$$

Here, for $\bar{\beta} < \frac{1}{4}$ we let

$$(6.10)\quad \omega^* = \frac{2}{1+\sqrt{1-4\bar{\beta}}}.$$

As shown in [19], the bounds for $S(\mathcal{S}_\omega)$ can be minimized by the following choices of ω

$$(6.11)\quad \omega_1 = \begin{cases} \dfrac{2}{1+\sqrt{1-2\bar{\mu}+4\bar{\beta}}} & \text{if } \bar{\mu} \le 4\bar{\beta} \\[2ex] \dfrac{2}{1+\sqrt{1-4\bar{\beta}}} & \text{if } \bar{\mu} > 4\bar{\beta}. \end{cases}$$

The corresponding values of $S(\mathcal{S}_{\omega_1})$ are

$$(6.12)\quad S(\mathcal{S}_{\omega_1}) \le \begin{cases} \dfrac{1 - \dfrac{1-\bar{\mu}}{\sqrt{1-2\bar{\mu}+4\bar{\beta}}}}{1 + \dfrac{1-\bar{\mu}}{\sqrt{1-2\bar{\mu}+4\bar{\beta}}}}, & \text{if } \bar{\mu} \le 4\bar{\beta} \\[3ex] \dfrac{1 - \sqrt{1-4\bar{\beta}}}{1 + \sqrt{1-4\bar{\beta}}}, & \text{if } \bar{\mu} > 4\bar{\beta}. \end{cases}$$

For the model problem it is easy to show that

$$(6.13)\quad S(B) = \cos \pi h, \qquad S(LU) \le \frac{1}{4} \cos^2 \frac{\pi h}{2} .$$

Letting $\bar{\mu} = \cos \pi h$ and $\bar{\beta} = \frac{1}{4} \cos^2(\pi h/2)$ we note that $\bar{\mu} < 4\bar{\beta} < 2\sqrt{\bar{\beta}}$ and

$$(6.14)\quad \omega_1 = \frac{2}{1 + \sqrt{3} \sin \frac{\pi h}{2}} .$$

Moreover,

$$(6.15)\quad S(\mathcal{S}_{\omega_1}) \le \frac{1 - \frac{2}{\sqrt{3}} \sin \frac{\pi h}{2}}{1 + \frac{2}{\sqrt{3}} \sin \frac{\pi h}{2}} \sim 1 - \frac{2\pi h}{\sqrt{3}} .$$

This is slightly greater than $S(\mathcal{L}_{\omega_b}) \sim 1 - 2\pi h$. However, the accelerated SSOR method gives much faster convergence than the SOR method. The reciprocal rate of convergence is

$$(6.16) \quad RR(\mathcal{S}_{\omega_1}) \sim \frac{\sqrt{3}}{2\pi} h^{-1}.$$

By (5.11), the reciprocal asymptotic average rate of convergence for the accelerated SSOR method is

$$(6.17) \quad RR_\infty(\mathcal{G}_n) \sim \frac{3^{\frac{1}{4}} h^{-\frac{1}{2}}}{2\sqrt{2}\sqrt{\pi}}$$

which is much less than $RR(\mathcal{L}_{\omega_b}) = (2\pi h)^{-1}$.

7. A BOUND FOR S(LU)

The choice of ω and the determination of a bound for $S(\mathcal{S}_\omega)$ require bounds $\bar{\mu}$ and $\bar{\beta}$ for $S(B)$ and $S(LU)$, respectively. We have already given a bound for $S(B)$, namely, (4.1). In [19] it is shown that

$$(7.1) \quad S(LU) \le \bar{\beta} = \max_{(x,y)\in R_h} \{\beta_3(x,y)[\beta_1(x-h,y) + \beta_2(x-h,y)] + \beta_4(x,y)[\beta_1(x,y-h) + \beta_2(x,y-h)]\},$$

where the functions $\beta_i(x,y)$, $i = 1,2,3,4$ are defined by (6.3). The result follows easily from the fact that $S(LU) \le \|LU\|_\infty = \bar{\beta}$.

It is also shown in [19] that

$$(7.2) \quad \bar{\beta} \le \frac{1}{4} + O(h^2)$$

provided that $A(x,y)$ and $C(x,y)$ belong to class $D^{(2)}$ in $R+S$. Moreover, by (4.1) it follows that

$$(7.3) \quad \bar{\mu} \le 1 - c_1 h^2 + O(h^4)$$

for some $c_1 > 0$. From (6.12) we have

$$(7.4) \quad S(\mathcal{S}_{\omega_1}) \le \begin{cases} \dfrac{1 - \sqrt{\frac{1-\bar{\mu}}{2}}\left[1 + \frac{2(\bar{\beta}-\frac{1}{4})}{1-\bar{\mu}}\right]^{-\frac{1}{2}}}{1 + \sqrt{\frac{1-\bar{\mu}}{2}}\left[1 + \frac{2(\bar{\beta}-\frac{1}{4})}{1-\bar{\mu}}\right]^{-\frac{1}{2}}}, & \text{if } \bar{\beta} \ge \frac{1}{4} \\[2ex] \dfrac{1 - \sqrt{\frac{1-\bar{\mu}}{2}}}{1 + \sqrt{\frac{1-\bar{\mu}}{2}}}, & \text{if } \bar{\beta} \le \frac{1}{4}, \end{cases}$$

where ω_1 is given by (6.11). Hence, since the ratio $(\bar{\beta}-\frac{1}{4})/(1-\bar{\mu})$ is bounded as $h \to 0$, it follows that

(7.5) $$S(\mathcal{S}_{\omega_1}) \leq 1 - c_2 h + o(h)$$

as $h \to 0$ for some positive c_2.

From (7.5) it follows that $RR(\mathcal{S}_{\omega_1}) = O(h^{-1})$ and hence, using acceleration, we can obtain a reciprocal asymptotic average convergence rate of $O(h^{-\frac{1}{2}})$.

8. A COMPUTATIONAL PROCEDURE

We now describe a computational procedure based on the accelerated SSOR method for solving a class of elliptic boundary value problems involving the differential equation (4.2).

1. Choice of starting vector. Let $u^{(0)} = 0$ or some vector u such that

(8.1) $$Q(u) = \tfrac{1}{2}(u,Au) - (b,u) < 0.$$

Such a vector will be closer to the true solution $\bar{u}$ than is the vector $u^{(0)} = 0$ in the sense that

(8.2) $$\|u-\bar{u}\|_{A^{\frac{1}{2}}} < \|0-\bar{u}\|_{A^{\frac{1}{2}}} = \|\bar{u}\|_{A^{\frac{1}{2}}} .$$

Here we define the $A^{\frac{1}{2}}$-norm of a vector v by

$$\|v\|_{A^{\frac{1}{2}}} = \|A^{\frac{1}{2}}v\|.$$

2. Choice of bounds for S(B) and S(LU). Compute the bound $\bar{\mu}$ for S(B) by (4.1) and the bound $\bar{\beta}$ for S(LU) by (7.1).

3. Choice of ω_1 and bound for $S(\mathcal{S}_{\omega_1})$. Compute ω_1 by (6.11) and a bound for $S(\mathcal{S}_{\omega_1})$ by (6.12).

4. Acceleration. With $\alpha = 0$ and $\beta = S(\mathcal{S}_{\omega_1})$ determine σ by (5.5) and r by (5.8).

4a. Variable extrapolation. Find m as the smallest integer such that

$$\text{(8.4)} \qquad \left[-\frac{1}{m}\log\frac{2r^{m/2}}{1+r^m}\right]^{-1} \le (.8)\,\frac{1}{-\frac{1}{2}\log r}\,.$$

(This choice of m guarantees that the reciprocal average rate of convergence for the variable extrapolation method does not exceed 125% of the reciprocal asymptotic rate of convergence of the semi-iterative method.)

Determine $\theta_1, \theta_2, \ldots, \theta_m$ by (5.16) and carry out the variable extrapolated SSOR method using (5.15).

4b. Semi-iteration (alternative to 4a). Carry out the semi-iterative method (5.2) with $\alpha = 0$, $\beta = S(\mathcal{S}_{\omega_1})$.

5. Convergence tests. The variable extrapolation method should be carried out for tm iterations where

$$\text{(8.5)} \qquad \left(\frac{2r^{m/2}}{1+r^m}\right)^t \le \zeta\,,$$

where ζ is a given quantity, say 10^{-6}. If this test is satisfied then, except for rounding errors

$$\text{(8.6)} \qquad \frac{\|u^{(tm)}-\bar{u}\|_{A^{\frac{1}{2}}}}{\|\bar{u}\|_{A^{\frac{1}{2}}}} \le \zeta\,.$$

The semi-iterative method is carried out until

$$\text{(8.7)} \qquad \frac{2r^{n/2}}{1+r^n} \le \zeta\,.$$

If this test is satisfied, then, except for rounding errors,

$$\text{(8.8)} \qquad \frac{\|u^{(n)}-\bar{u}\|_{A^{\frac{1}{2}}}}{\|\bar{u}\|_{A^{\frac{1}{2}}}} \le \zeta\,.$$

9. NUMERICAL RESULTS

We now describe some numerical experiments which were carried out using the accelerated SSOR method. The differential equation (4.2) was solved in the unit square with boundary values zero on all sides except where y = 0. On the side y = 0, the values of u(x,y) were unity. Various choices of the coefficient functions A(x,y) and C(x,y) were used. (See Concus and Golub [3] and Axelsson [1].) In each case variable extrapolation and semi-iteration were used. Values of ω based on estimates of $\bar{\mu}$ and $\bar{\beta}$ determined by (4.1) and (7.1), respectively, were used. Also, values of ω were used which were determined to be optimum. (Here, values of $S(\mathcal{S}_\omega)$ were determined for various values of ω using the power method.) For comparison, results using the SOR method with the optimum ω are given. The numbers of iterations given refer to double sweeps for the SSOR method and to single sweeps for the SOR method.)

The following tentative conclusions are made.

(1) The number of iterations required for convergence using the extrapolated SSOR method varies as $h^{-\frac{1}{2}}$ as compared to h^{-1} for the SOR method. The number of iterations required by the former method is much less than for the SOR method when the mesh size h is small.

(2) The estimated value of ω is reasonably good. However, a significant saving can be realized by using the optimum ω instead of the estimated ω. A method which would dynamically improve the ω without "wasting" iterations would be worthwhile.

(3) If there is sufficient memory available to do so, the semi-iterative method is to be preferred to the variable extrapolated procedure. (However, the author is currently working on a dynamic procedure for parameter determination which requires the use of the variable extrapolation method instead of the semi-iterative method.)

10. SUMMARY AND CONCLUSION

Following the procedure outlined in Section 8, the accelerated SSOR method affords a potentially effective method for solving a large class of elliptic boundary value problems. It appears that the method has so far not received as much usage as it should.

11. ACKNOWLEDGEMENT

The author wishes to acknowledge the contribution of Vitalius Benkraitis of The University of Texas at Austin both in carrying out the numerical studies described above and in making useful suggestions concerning the theory. The cooperation of the University of Texas Computation Center in making its facilities available

Table 9.1 Numerical Results

		Estimated		Optimum		
	h	SSOR-VE	SSOR-SI	SSOR-VE	SSOR-SI	SOR
$A = C = 1$	1/20	25	19	20	16	44
	1/40	35	26	30	23	88
	1/80	45	37	40	32	174
$A = C = e^{10(x+y)}$	1/20	12	10	12	10	24
	1/40	20	15	16	14	48
	1/80	25	21	25	20	119
$A = \frac{1}{1+2x^2+y^2}$	1/20	35	28	20	17	45
$CC = \frac{1}{1+x^2+2y^2}$	1/40	50	40	30	23	90
	1/80	70	57	40	33	177
$A = C = \begin{cases} 1+x, & 0 \leqq x \leqq \frac{1}{2} \\ 2-x, & \frac{1}{2} \leqq x \leqq 1 \end{cases}$	1/20	24	21	20	17	46
	1/40	40	32	30	24	92
	1/80	60	49	40	33	180
$A = 1 + 4\|x - \frac{1}{2}\|^2$	1/20	35	28	25	19	43
$C = \begin{cases} 1, & 0 \leqq x < \frac{1}{2} \\ 9, & \frac{1}{2} \leqq x \leqq 1 \end{cases}$	1/40	50	40	30	25	86
	1/80	70	56	45	34	164
$A = 1 + \sin \frac{\pi(x+y)}{2}$	1/20	12	11	12	10	24
$C = e^{10(x+y)}$	1/40	20	15	20	15	47
	1/80	30	22	25	21	120

for the numerical work is also acknowledged.

12. BIBLIOGRAPHY

1. Axelsson, O. "A generalized SSOR method," BIT 13, 443-467 (1972).

2. Carré, B. A. "The determination of the optimum accelerating factor for successive over-relaxation," The Computer Journal 4, 73-78 (1961).

3. Concus, Paul, and Gene H. Golub. "Use of fast direct methods for the efficient numerical solution of nonseparable elliptic equations," Report STAN-CS-72-278, Computer Science Department, Stanford University, April 1972.

4. Golub, G. H. "The use of Chebyshev matrix polynomials in the iterative solution of linear systems compared with the method of successive overrelaxation," doctoral thesis, University of Illinois, Urbana, Illinois (1959).

5. Golub, G. H., and R. S. Varga. "Chebyshev semi-iterative methods, successive, overrelaxation iterative methods, and second-order Richardson iterative methods," Numer. Math., Parts I and II, 3, 147-168 (1961).

6. Hageman, L. A., and R. B. Kellogg, "Estimating optimum over-relaxation parameters," Math. Comp. 22, 60-68 (1968).

7. Kahan, W. "Gauss-Seidel methods of solving large systems of linear equations," doctoral thesis, University of Toronto, Toronto, Canada (1958).

8. Kulsrud, H. E. "A practical technique for the determination of the optimum relaxation factor of the successive over-relaxation method," Comm. Assoc. Comput. Mach. 4, 184-187 (1961).

9. Reid, J. K. "A method for finding the optimum successive over-relaxation parameter," The Computer Journal 9, 200-204 (1966).

10. Richardson, L. F. "The approximate arithmetical solution by finite differences of physical problems involving differential equations with an application to the stresses in a masonry dam," Philos. Trans. Roy. Soc. London Ser. A 210, 307-357 (1910).

11. Rigler, A. K. "Estimation of the successive over-relaxation factor," Math. of Comp. 19, 302-307 (1965).

12. Sheldon, J. "On the numerical solution of elliptic difference equations," Math. Tables Aids Comput. 9, 101-112 (1955).

13. Wachspress, E. L. "Iterative Solution of Elliptic Systems and Applications to the Neutron Diffusion Equations of Reactor Physics," Prentice-Hall, Englewood Cliffs, New Jersey (1966).

14. Young, David M. "Ordvac solutions of the Dirichlet problem," J. Assoc. Comput. Mach. 2, 137-161 (1955).

15. ______. "On the solution of linear systems by iteration," Proc. Sixth Symp. in Appl. Math. Amer. Math. Soc. VI, McGraw-Hill, New York (1956).

16. ______. "Iterative Solution of Large Linear Systems," Academic Press, New York (1971).

17. ______. "A bound on the optimum relaxation factor for the successive overrelaxation method," Numer. Math. 16, 408-413 (1971).

18. ______. "Second-degree iterative methods for the solution of large linear systems," J. Approx. Theory 25, 137-148 (1972).

19. ______. "On the accelerated SSOR method for solving large linear systems," to appear.

SOLUTION OF NONLINEAR SYSTEMS OF EQUATIONS

L.B. Rall

Mathematics Research Center,
University of Wisconsin.

1. Theoretical and practical difficulties

In the attempt to obtain a numerical solution of a partial differential equation, one is usually lead to consider a finite system of equations,

$$p_i(x_1,x_2,\ldots,x_n) = \qquad i=1,2,\ldots,n \tag{1.1}$$

These systems are commonly obtained by <u>collocation</u>, in which one assumes that the unknowns $x_1,x_2,\ldots,x_n$ are approximate values of the solution of the partial differential equation at points $P_1,P_2,\ldots,P_n$ in the region considered, or by <u>projection</u>, in which $x_1,x_2,\ldots,x_n$ are the coefficients of the expansion of an approximate solution ϕ of the partial differential equation in terms of some suitably chosen <u>basis functions</u> $\phi_1,\phi_2,\ldots,\phi_n$; that is,

$$\phi = x_1\phi_1 + x_2\phi_2 + \ldots + x_n\phi_n \tag{1.2}$$

Independently of how the system (1.1) is constructed, further progress on the original problem of obtaining a numerical solution of the given partial differential equation awaits the determination of $x_1,x_2,\ldots,x_n$. This may present theoretical as well as practical difficulties.

From the standpoint of theory, one is interested in finding out which of the alternatives,

$$\left.\begin{array}{ll}\text{(i)} & \text{existence and uniqueness,} \\ & \qquad\text{or} \\ \text{(ii)} & \text{nonexistence or nonuniqueness,}\end{array}\right\} \quad (1.3)$$

holds for <u>solution-vectors</u> $x^*=(x_1^*,x_2^*,\ldots,x_n^*)$ of the system (1.1). In the case of existence but nonuniqueness, a satisfactory theory should also specify the set X^* of solution-vectors x^*.

The alternative structure (1.3) of a theory for a system of equations has been completely elucidated for <u>linear</u> systems (MacDuffee, 1954), as would arise in the present context from linear partial differential equations. Even so, the actual solution of linear systems, particularly those of large size originating from partial differential equations, presents substantial practical problems. These have been dealt with elsewhere in this volume by Young (1973), and will not be dwelt upon further. In the nonlinear case, one does not have the comfort of a complete theory in general, the outstanding exceptions being polynomial equations in a single variable,

$$a_0z^n + a_1z^{n-1} +\ldots+ a_{n-1}z+a_n = 0 \qquad (1.4)$$

(MacDuffee, 1954), and the <u>algebraic eigenvalue problem</u>,

$$\left.\begin{array}{l} a_{11}x_1+a_{12}x_2 +\ldots+ a_{1n}x_n = \lambda x_1, \\ a_{21}x_1+a_{22}x_2 +\ldots+ a_{2n}x_n = \lambda x_2, \\ \ldots \quad \ldots \quad \ldots \quad \ldots \quad \ldots \\ a_{n1}x_1+a_{n2}x_2 +\ldots+ a_{nn}x_n =\lambda x_n, \\ \\ x_1^2+x_2^2 +\ldots+ x_n^2 = 1, \end{array}\right\} \qquad (1.5)$$

which is a system of n+1 quadratic equations for the n+1 unknowns $\lambda, x_1,x_2,\ldots,x_n$ (Wilkinson, 1965). Even in these special cases, one is confronted with practical difficulties (Dejon and Henrici, 1969; Wilkinson, 1965). However, in spite of these considerations, one may regard the theory of linear systems (1.1) and the nonlinear problems (1.4) and (1.5) to be settled.

For general nonlinear systems, the guidance of a complete theory is missing. In its place, one looks for a computational theory, which usually provides conditions under which alternative (1.3i) holds, and prescribes a sequence $\{x^{(k)}\}$ of vectors $x^{(k)} = (x_1^{(k)}, x_2^{(k)}, \ldots, x_n^{(k)})$ which converges accordingly to the solution vector x , starting from some suitable initial vector $x^{(0)} = (x_1^{(0)}, x_2^{(0)}, \ldots, x_n^{(0)})$. Computational theories may be classified into three categories: (i) global; (ii) semi-local; and (iii) local. A global theory asserts the existence and uniqueness of x^* and the convergence of the sequence $\{x^{(k)}\}$ to x^* independently of the choice of $x^{(0)}$. A semi-local theory includes conditions on $x^{(0)}$, and asserts existence and uniqueness of x^* only in some neighborhood U of $x^{(0)}$, with U containing the sequence $\{x^{(k)}\}$ which converges to x^*. A local theory, on the other hand, assumes the existence of x^*, and asserts the convergence of $\{x^{(k)}\}$ to x^* for initial vectors $x^{(0)}$ sufficiently close to x^*, and often gives information concerning the rate of convergence. Computational theories, although lacking completeness, are useful in practice. Examples of the various types of computational theories will be given below. Once again, practical difficulties arise in the application of computational theories, which give rise to the need for an error analysis as part of an actual computational method for the solution of systems of equations (Rall, 1965). There are several types of error to be considered.

(i) Truncation error, which arises from the fact that one computes only a finite number of terms of the sequence $\{x^{(k)}\}$, and takes the final element obtained as the approximate value of the solution-vector x^*. This type of error is fairly well understood from a theoretical standpoint, and most theorems which assert convergence of a computational method also give estimates for the truncation error.

(ii) Round-off error, due to the fact that the elements $x^{(1)}, x^{(2)}, \ldots, x^{(k)}$ of the sequence converging to x^* are themselves not calculated exactly. One way to approach this error is by a forward error analysis, in which computational errors at each step are accumulated to give an estimate of the round-off error in the final result. This and a bound for the truncation error can then be used to estimate the accuracy of the computed vector as an approximation to x^*. The results obtained in this way can overestimate the actual error grossly. An automatic method for forward error analysis is provided by the use of the techniques of interval analysis (Moore, 1966; Hansen, 1969). In certain cases, interval analysis allows one to obtain satisfactory estimates of round-off error with little effort, assuming that the necessary computer programs. There is also backward error analysis, which assumes that the vectors computed are the exact results of a perturbed process, and estimates the magnitude of the perturbation. This has been highly developed for linear systems, single polynomial equations, and the algebraic eigenvalue problem (Wilkinson, 1963, 1965), but has been exploited relatively little for nonlinear problems of various other types.

(iii) Errors in coefficients of the given system may arise, due to uncertainty in physical measurements and other circumstances. In this case, the original equation (1.1) may be considered to belong to a family of perturbed systems

$$p_i(x_1,x_2,\ldots,x_n)+q_i(x_1,x_2,\ldots,x_n)=0 \quad i=1,2,\ldots,n, \tag{1.6}$$

and one would want to estimate the error resulting from solving (1.1) instead of (1.6) for perturbation functions $q_i(x)$ of given form and magnitude.

All of these sources of error may be dealt with by considering the computed result to be a new initial approximation $x^{(0)}$ and x^*. This is called a posteriori error estimation. (Interval analysis is frequently useful for this purpose (Rockne and Lancaster,1969); Kuba and Rall, 1972). One can also consider a backward form of a posteriori error analysis, in which the result form of *a posteriori* error analysis, in which the result obtained is considered to be the exact solution of a perturbed system (1.6), and the magnitudes of the functions $q_i(x)$ can be used to construct a measure of the error.

In addition to the theoretical problems of the existence of x^*, finding a computational method which can be shown to converge to x^*, and error estimation, there are practical difficulties to be overcome, even when no theoretical barrier is present. These problems arise usually from the *size* and *complexity* of the given system. The *size* of the system is simply the number n of equations and unknowns; this is also called the *order* of the system. Most useful approximations of partial differential equations by systems of equations *converge* in the sense that the solutions of the system give as accurate solutions of the partial differential equation as one pleases, provided that n is large enough. However, to reduce this *discretization error* sufficiently, one may be confronted with a system which is too large to handle with the computational time and facilities available. One would want a solution method which would work for a given approximation method more or less independently of n, at least until round-off error or time and space limitations would render further increase in n impractical.

Another practical problem is the *complexity* of the system (1.1), which will be taken to mean a measure of the effort required in terms of time or arithmetic operations to evaluate the functions $p_i(x)$, $i=1,2,...,n$. Approximations to partial differential equations which are more accurate for a given value of n generally lead to systems of greater complexity. It will be seen later that an exchange between size and complexity is a fairly general phenomenon in the nonlinear case. The decision to use a more accurate approximation method, such as one based on reformulation of the partial differential as an integral equation, will once again depend on the availability of solution methods and adequate computational power.

As in other cases, once the types of nonlinear systems which arise in the approximate solution of nonlinear partial differential equations of practical importance have been identified and isolated, then special techniques for their solution can be devised.

These special techniques will usually be adapted from more general procedures, and will depend heavily on the particular structure of the given type of system. Consequently, attention will be devoted in what follows to some methods which have been found to be effective in practice for the solution of more or less general systems, with possible adaptations to systems arising from partial differential equations being mentioned wherever appropriate. Even this limited goal is beyond the scope of a paper of reasonable size; however, one can find a much broader treatment of the subject in the books on nonlinear equations and systems by Ostrowski (1966), Ortega and Rheinboldt (1970), Rabinowitz (1970), and the books on the solution of nonlinear operator equations by Rall (1969), and the Russian quintet of Krasnosel'skii, Vainikko, Zabreiko, Rutitskii, and Stetsenko (1972). The book by Ortega and Rheinboldt includes an extensive bibliography.

2. Direct methods for nonlinear systems

A method for solving the system (1.1) will be said to be <u>direct</u> if it reduces the problem to the solution of an equivalent single equation

$$f(c) = 0 \tag{2.1}$$

in one unknown; that is, relationships

$$\left.\begin{aligned} &x_i = x_i(c), \quad i=1,2,\ldots,n, \\ &c = c(x_1, x_2, \ldots, x_n) \end{aligned}\right\} \tag{2.2}$$

exist such that if $c=c^*$ is a solution of (2.1), then $x_1^*=x_1(c^*)$, $x_2^*=x_2(c^*),\ldots,x_n^*=x_n(c^*)$ satisfy (1.1), and also $c^*=c(x_1^*,x_2^*,\ldots,x_n^*)$ is a solution of (2.1) for any solution-vector x of (1.1). Although the solution of (2.1) may present difficult problems, single scalar equations are better understood in theory and practice than systems in general; see, for example, the books by Durand (1960), Householder (1970), Ostrowski (1966), Rabinowitz (1970), Traub (1964) Zaguskin (1961).

The most obvious direct method is <u>elimination</u>. Supposing, for example, that the nth equation of (1.1) may be solved explicitly for x_n,

$$x_n = g_n(x_1,x_2,\ldots,x_{n-1}), \tag{2.3}$$

then this result may be substituted into the first n-1 equations to obtain the smaller system

$$p_i^{(1)}(x_1,x_2,\ldots,x_{n-1})=0, \quad i=1,2,\ldots,n-1 . \tag{2.4}$$

If this process can be continued, then one eventually arrives at the single scalar equation

$$p_1^{(n-1)}(x_1) = 0, \tag{2.5}$$

in which the unknown x_1 plays the part of c in (2.1).

In the case of linear systems, elimination presents little conceptual or practical difficulty; the intermediate systems and the final equation (2.5), if it is obtained, are all linear, and there is no increase in complexity. For nonlinear systems, elimination may increase complexity to the point of being useless from a practical standpoint. In addition, there is the possibility that the solution of a given equation for some unknown, as indicated by (2.3), is not unique, so that one must consider a number of reduced systems to preserve equivalence. Of course, if one is only interested in obtaining some solutions of the original system, rather than all of them, then it is only necessary to insure that solutions of the reduced system at each step will provide solutions of the previous system.

On the surface, elimination does not appear to be a promising method for nonlinear systems. However, in a given situation, its use should be investigated, as even a partial elimination of selected unknowns might result in a helpful decrease in the size of the system without a drastic increase in complexity.

Some nonlinear problems have a structure which lends itself immediately to treatment by direct methods. For example, suppose that the kernel G(s,t) of the Hammerstein integral equation,

$$x(s)- \int_0^1 G(s,t)f(t,x(t))dt=0, \tag{2.6}$$

is a Green's function of the form

$$(u(s)v(t), \quad 0 \quad t \quad s,$$

$$G(s,t) = \begin{cases} u(s)v(t), & 0 \le t \le s, \\ u(t)v(s), & s \le t \le 1, \end{cases} \tag{2.7}$$

Then (Rall, 1973a), equation (2.6) is equivalent to the system

$$\left.\begin{aligned} x(s) - \int_0^s K(s,t)f(t,x(t))dt &= cv(s), \\ c = \int_0^1 u(t)f(t,x(t))dt, & \end{aligned}\right\} \tag{2.8}$$

where

$$K(s,t)=u(s)v(t)-u(t)v(s). \tag{2.9}$$

Solving the Volterra equation in (2.8) for $x(s)=x(s;c)$ and substituting the result into the second equation gives a single scalar equation

$$c = \Phi(c) . \tag{2.10}$$

This direct method for the nonlinear integral equation (2.6) also applies to the nonlinear system obtained from (2.8) by the use of numerical integration. For example, the trapezoidal rule with $h=\frac{1}{n}$ gives the system

$$\left.\begin{aligned} x_0 &= cv_0, \\ x_1 &= cv_1 + \tfrac{1}{2}hK(h,0)f(0,x_0), \\ x_k &= cv_k + \tfrac{1}{2}hK(kh,0)f(0,x_0) + \sum_{i=1}^{k-1} K(kh,ih)f(ih,x_i)) \\ c &= \tfrac{1}{2}hu_0 f(0,x_0) + h\sum_{i=1}^{n-1} u_i f(ih,x_i) + \tfrac{1}{2}hu_n f(1,x_n)) \end{aligned}\right\} \tag{2.11}$$

as $K(s,s)=0$, where $v_i=v(ih)$, $u_i=u(ih)$, and x_i is the desired approximation to $x(ih)$, $i=0,1,\ldots,n$. Because of the lower triangular structure of (2.11), one can in principle solve it explicitly for $x_0,x_1,\ldots,x_n$ in terms of c, and substitute the results into the last equation to obtain (2.10). If $f(t,x(t))$ is a polynomial in $x(t)$, as in the <u>Duffing</u>

equation (Duffing, 1918), where

$$\left.\begin{aligned} f(t,x(t)) &= \alpha\left(x(t) - \frac{x^3(t)}{3!}\right) + \beta\sin t, \\ K(s,t) &= s-t, \qquad v(s) = s, \end{aligned}\right\} \quad (2.12)$$

then (2.10) will be a polynomial equation for c. In actual practice, one would use the system (2.11) to evaluate $\Phi(c)$, and perhaps also its derivatives.

The above example, which arises from a boundary value problem for an ordinary differential equation, suggests that reformulation of certain partial differential equations as integral equations may lead to direct methods for their numerical solution. It should also be noted that the numerical integration of the system (2.8) instead of the original equation (2.6) avoids having to deal with the singularity in the derivative of the Green's function G(s,t) at x=t.

3. Useful concepts from functional analysis

From the standpoint of functional analysis, that is, analysis on normed linear spaces, the system (1.1) is a special case of the operator equation

$$P(x) = 0 \qquad (3.1)$$

where the operator P maps a Banach space X (or a subset D of X, called the domain of P) into a Banach space Y. In the case of present interest, X and Y are both n-dimensional vector spaces. This identification allows the techniques for the solution of general operator equations to be applied to the system (1.1). Although the results obtained can in almost all cases be derived on the basis of classical analysis, the use of functional analysis permits a conceptual and notational simplicity of treatment. In addition, the original partial differential equation can also be considered to be an operator equation of the form (3.1) in an appropriate function space. This allows one to develop strict analogies between numerical methods for the solution of the approximating system and analytic procedures for solving the partial differential equation, and makes it easier to interpret the numerical solution as an approximation to the desired function.

The concepts of functional analysis which will be most useful in the following include:

(i) vector norms;
(ii) linear, multilinear, and power operators;
(iii) operator norms;
(iv) derivatives;
(v) integrals;
(vi) Taylor's theorem and power series;
(vii) scalar majorants.

Most of these topics have been discussed adequately in the literature for the present purposes (Taylor, 1958); Dieudonne, 1960; Rall, 1969; Ortega and Rheinboldt, 1970). A brief summary will be given here of the more important details.

The _norm_ of a vector x, denoted by $||x||$, is a non-negative real function of x (a real-valued _functional_ in the usual terminology), which has the same properties as the absolute value of a number. A useful norm for numerical analysis is, for example,

$$||x|| = \max_{(i)} |x_i|. \tag{3.1}$$

There are many other possible definitions. For example, one might wish to _scale_ the variables $x_1, x_2, \ldots, x_n$ by positive constants $a_1, a_2, \ldots, a_n$, and use the norm

$$||x|| = \sum_{i=1}^{n} |a_i x_i|, \tag{3.2}$$

and so on. All n-dimensional real or complex normed linear vector spaces consist of the same set of elements, but may be normed differently for convenience, and will be denoted by X,Y,... . Ordinarily, the symbol $|| \; ||$ will be used for the norm in various spaces, it being clear from the context to which space the argument belongs. In case a distinction is necessary, $|| \; ||_X$, $|| \; ||_Y, \ldots$ will be used. An important property of n-dimensional spaces is the _equivalence_ of norms for them (Taylor, 1958). That is, if X and Y are both n-dimensional, then positive constants a,b exist such that

$$a||z||_X \leq ||z||_Y \leq b||z||_X \qquad (3.3)$$

for all n-dimensional vectors z. Convergence of a sequence of vectors in one norm thus implies the convergence of the sequence in any other norm.

A linear operator A from nn n-dimensional space X into an n-dimensional space Y can be represented uniquely by an m×n <u>matrix</u>

$$A=(a_{ij}) = \begin{pmatrix} a_{11} & a_{12} & \cdots & a_{1n} \\ a_{21} & a_{22} & \cdots & a_{2n} \\ \cdots & \cdots & \cdots & \cdots \\ a_{m1} & a_{m2} & \cdots & a_{mn} \end{pmatrix}, \qquad (3.4)$$

with constant coefficients a_{ij}, $i=1,2,\ldots,m$; $j=1,2,\ldots,n$ (Taylor, 1958). The components y_i of the vector $y=Ax$ are given by the well-known <u>row-by-column</u> <u>rule</u>,

$$y_i = \sum_{j=1}^{n} a_{ij}x_j, \qquad i=1,2,\ldots,m. \qquad (3.5)$$

These matrices form an mn-dimensional linear space XY for the natural definitions of addition and scalar multiplication. This provides the possibility of defining linear operators from X to XY, which will be called <u>bilinear</u> operators from X into Y. A bilinear operator B has the representation of an m×n×n array

$$B=(b_{ijk}), \quad i=1,2,\ldots,m; \quad j,k=1,2,\ldots,n. \qquad (3.6)$$

The components a_{ij} of the matrix $A=Bx$ are given by

$$a_{ij} = \sum_{k=1}^{n} b_{ijk}x_k, \quad i=1,2,\ldots,m; \; j=1,2,\ldots,n, \qquad (3.7)$$

by analogy to (3.5). The linear operator Bx from X into Y may, of course, operate on a second point z of X to yield the vector

$$y=(Bx)z = Bxz, \qquad (3.8)$$

with components

$$y_i = \sum_{j=1}^{n} \sum_{k=1}^{n} b_{ijk} x_k z_j, \quad i=1,2,\ldots,m. \tag{3.9}$$

This transformation is linear with respect to each of the vector arguments x and z separately; this motivates the nomenclature "bilinear". A bilinear operator is symmetric if

$$b_{ijk} = b_{ikj}, \quad i=1,2,\ldots,m;\ j,k=1,2,\ldots,n. \tag{3.10}$$

for a symmetric operator, Bxz=Bzx. The operator $\overline{B}=\overline{b}_{ijk})$ with components

$$\overline{b}_{ijk} = \frac{1}{2}(b_{ijk}+b_{ikj}),\ i=1,2,\ldots,m;\ k=1,2,\ldots,n, \tag{3.11}$$

is always symmetric, and is called the average of B. Bilinear operators from X into Y also form a linear space, which will be denoted by X^2Y.

By induction, the above concepts can be extended to multilinear operators of higher order. A d-linear operator, or multilinear operator of order d, may be represented by the array

$$D=(d_{ij_1\ldots j_d}),\ i=1,2,\ldots,m;\ j_k=1,2,\ldots,n;\ k=1,2,\ldots,d, \tag{3.12}$$

and will be a linear operator from X into the space $X^{d-1}Y$ of (d-1)-linear operators from X into Y. By extending (3.9), the operation of D on d points of X to give a vector $y=Dx^{(1)}\ldots x^{(d)}$ of Y can be specified. D will be symmetric if it does not change value for arbitrary permutations of the subscripts $j_1,\ldots,j_d$. The average $\overline{D}$ of D is the symmetric d-linear operator obtained by averaging over all d permutations of D obtained by permuting $j_1,\ldots,j_d$ (Rall, 1961).

The introduction of multilinear operators permits the definition of some simple nonlinear operators from X into Y called power operators. The quadratic transformation Bxx, B bilinear, is a simple generalization of the square of a scalar variable, and so on. In general, for D d-linear,

$$Dx^d = D\overbrace{x\ldots x}^{d} \tag{3.13}$$

defines a power operator of degree d on x. Linear combinations of such operators with linear operators and constant vectors are called polynomial operators. In (3.13), the coefficient operator D may be assumed to be symmetric without loss of generality, for it may be replaced by $\bar{D}$ without altering the value of the transformation (Rall, 1961).

The spaces of linear and multilinear operators from X into Y will be assumed to be normed in a special fashion, by means of the operator norms induced by the norms assigned to X and Y. If A is a linear operator from X into Y, then

$$||A|| = \max_{||x||=1} ||Ax|| \qquad (3.14)$$

is the operator norm for XY. One has immediately

$$||Ax|| \leq ||A||.||x|| \qquad (3.15)$$

for all x, using the operator norm of A. As an example of an operator norm, if X and Y both have the norm (3.1), then

$$||A|| = \max_{(i)} \sum_{j=1}^{n} |a_{ij}| \qquad (3.16)$$

By induction, one can extend operator norms to multilinear operators, and obtain inequalities analogous to (3. 5). If $||B||$ denotes the operator norm of a bilinear operator B as a linear operator from X into XY, then, using the fact that $||Bx||$ is the operator norm for Bx and inequality (3.15), one has

$$||Bxz|| \leq ||B||.||x||.||z||, \qquad (3.17)$$

for all x,z in X. For the norm (3.1),

$$||B|| \leq \max_{(i)} \sum_{j=1}^{n} \sum_{k=1}^{n} |b_{ijk}|, \qquad (3. 18)$$

but equality is not necessarily attained. Operator norms for multilinear operators (or upper bounds for them) will be useful in the following in connection with the construction of scalar majorant functions.

The definition of the derivative to be employed here is the one due to Fréchet (Rall, 1969). Other useful derivatives exist under somewhat less restrictive conditions, such as the one due to Gâteaux (Ortega and Rheinboldt, 1970), but the Fréchet derivative is usually adequate from a conceptual and computational standpoint. The operator P from X into Y is said to be _differentiable at_ x if a linear operator $P'(x)$ from X into Y exists such that

$$\lim_{||\Delta x|| \to 0} \frac{||P(x+\Delta x)-P(x)-P'(x)\Delta x||}{||\Delta x||} = 0 \tag{3.19}$$

If $P'(x)$ exists, then it is represented by the _Jacobian matrix_ with elements

$$P'(x)_{ij} = \frac{\partial p_i(x)}{\partial x_j}, \quad i=1,2,\ldots,m;\ j=1,2,\ldots,n. \tag{3.20}$$

Higher derivatives are defined by repeated application of (3.19). If the (d-1)st derivative of P exists, then $P^{(d)}(x)$ will have to satisfy the condition

$$\lim_{||\Delta x|| \to 0} \frac{||P^{(d-1)}(x+\Delta x)-P^{(d-1)}(x)-P^{(d)}(x)\Delta x||}{||\Delta x||} = 0 \tag{3.21}$$

It follows by induction that if $P^{(d)}(x)$ exists, then it will be a d-linear operator, and it is also known that it will be symmetric (Rall, 1969). For example, the second derivative is represented by the _Hessian operator_ with components

$$P''(x)_{ijk} = \frac{\partial^2 p_i(x)}{\partial x_k \partial x_j}, \quad i=1,2,\ldots,m;\ j,k=1,2,\ldots,n, \tag{3.22}$$

with similar expressions for derivatives of higher order.

Integrals may also be defined by simple generalization of the Riemann integral of ordinary analysis (Rall, 1969). Given points $x^{(0)}$, $x^{(1)}$ of X, one defines the <u>line segment</u> $x^{(0)}, x^{(1)}$ from $x^{(0)}$ to $x^{(1)}$ to be the set

$$[x^{(0)},x^{(1)}]=\{x(\lambda):\ x(\lambda)=(1-\lambda)x^{(0)}+\lambda x^{(1)},\ 0\leq\lambda\leq 1\} \tag{3.23}$$

By partitioning the real interval [0,1] and forming Riemann sums at the corresponding points of $[x^{(0)},x^{(1)}]$, and taking $||x(\lambda_{i+1})-x(\lambda_i)||$ as the length of the ith subinterval, one arrives at the definition

$$\int_{x^{(0)}}^{x^{(1)}} p(x)dx = \int_0^1 P((1-\lambda)x^{(0)}+\lambda x^{(1)})d\lambda, \tag{3.24}$$

provided that the Riemann sums converge. If the operator P is differentiable the required number of times, then one has Taylor's theorem (Rall, 1969),

$$\left.\begin{aligned} P(x+z)=P(x)+\sum_{k=1}^{d-1}\frac{1}{k!}P^{(k)}(x)z^k + \\ +\int_0^1 P^{(d)}(x+\lambda z)z^d\,\frac{(1-\lambda)^{d-1}}{(d-1)!}\,d\lambda \end{aligned}\right) \tag{3.25}$$

As in scalar calculus, it follows from (3.25) that if $P^{(d)}$ satisfies a <u>Lipschitz condition</u>

$$||P^{(d)}(x^{(1)})-P^{(d)}(x^{(0)})||\leq K||x^{(1)}-x^{(0)}|| \tag{3.26}$$

for $x^{(0)}$, $x^{(1)}$ in [x,x+z], then

$$||P(x+z)-\sum_{k=0}^{d}\frac{1}{k!}P^{(k)}(x)z^k||\leq\frac{K||z||^{d+1}}{(d+1)!} \tag{3.27}$$

Inequality (3.27) also holds if $P^{(d+1)}$ exists and $||P^{(d+1)}(x+\lambda z)|| \leq K$, $0 \leq \lambda \leq 1$. For operators P having derivatives of all orders, the above considerations suggest the possibility of a <u>Taylor series expansion</u>

$$P(x+z) = \sum_{k=0}^{\infty} \frac{1}{k!} P^{(k)}(x)z^k, \tag{3.28}$$

where, of course, $\frac{1}{0!} P^{(0)}(x) = P(x)$ by definition. The following ideas are useful in discussion of the convergence of expressions of the form (3.28).

Given an operator P from X into Y, a real-valued function p(r) of a single variable r is called a <u>scalar majorant function</u> for P if

$$||P(x)|| \leq p(||x||). \tag{3.29}$$

For example, if P is continuous, then

$$p(r) = \max_{||x||=r} ||P(x)|| \tag{3.30}$$

exists, but may be difficult to calculate. For P defined by a <u>power series</u>,

$$P(x) = A_0 + A_1 x + A_2 x^2 + \ldots, \tag{3.31}$$

where A_0 is a vector, A_1 is a linear operator, A_2 is bilinear, and so on, suppose that constants a_k are known such that

$$a_k \geq ||A_k||, \quad k=0,1,2,\ldots . \tag{3.32}$$

The series

$$p(r) = a_0 + a_1 r + a_2 r^2 + \ldots \tag{3.33}$$

is called a <u>scalar majorant series</u> for (3.31). If (3.33) converges for $|r|<R$, $R>0$, then (3.31) evidently converges uniformly for $||x||<R$, and is in fact analytic. In case that $R=\infty$, (3.31) is called an <u>entire function</u> of x. A simple example of an entire function is a <u>polynomial</u> of degree d,

$$P_d(x)=A_0+A_1x+\ldots+A_dx^d, \tag{3.34}$$

in which the expansion (3.31) is finite. Corresponding to (3.33), one has the <u>scalar majorant polynomial</u>

$$p_d(r)=a_0+a_1r+\ldots+a_dr^d. \tag{3.35}$$

In many cases, it will be useful to use scalar majorant functions to settle questions relating to the existence of solutions and the convergence of numerical methods. This methodology applies to nonlinear operator equations in general, and thus to nonlinear partial differential equations as well as nonlinear systems of equations.

4. <u>Quadratic and polynomial systems of equations</u>

The introduction of multilinear operators and polynomials leads directly to the consideration of the simple class of <u>algebraic</u> (or <u>polynomial</u>) systems

$$P_d(x)\equiv A_dx^d+\ldots+A_2x^2+A_1x+A_0=0 \tag{4.1}$$

of degree $d\geq 2$. In (4.1), it will be assumed that the multilinear coefficient operators $A_2, A_3, \ldots, A_d$ are all symmetric. The simplest nonlinear algebraic systems are the <u>quadratic</u> systems $(d=2)$,

$$Q(x)\equiv A_2x^2+A_1x+A_0=0. \tag{4.2}$$

Quadratic systems arise directly from certain partial differential equations, for example, by applying finite-difference techniques to the equation

$$\Delta u = u^2, \tag{4.3}$$

(Greenspan, 1965; Rall, 1969), or from a similar approximation of the Navier-Stokes equations for incompressible flow. It will be shown later that quadratic systems are also the most general polynomial systems in the sense that any system (4.1) may be transformed into an equivalent quadratic system. This decrease in complexity, as one might expect, is accompanied by an increase in size.

The application of analytic methods to polynomial systems is assisted by the fact that many operations on polynomial operators follow exactly the same rules as for the manipulation of scalar polynomials. For example, the derivative $P'_d(x)$ of the polynomial operator P_d at x is the linear operator

$$P'_d(x)=dA_d x^{d-1} +\ldots+ 2A_2 x+A_1 \tag{4.4}$$

and so on. Furthermore, if

$$p_d(r)=a_d r^d +\ldots+ a_2 r^2+a_1 r+a_0 \tag{4.5}$$

is a scalar majorant polynomial for $P_d(x)$, then

$$p'_d(r)=da_d r^{d-1} +\ldots+2a_2 r+a_1 \tag{4.6}$$

is a scalar majorant polynomial for $P'_d(x)$; in general, $p_d^{(k)}(r)$ will be a scalar majorant polynomial for $P_d^{(k)}(x)$.

In dealing with polynomial systems, it is also frequently useful to change the origin of the space to some given point x. One has

$$P_d(x+z)=P_d(x)+P'_d(x)z+\ldots+ \frac{1}{d!} P_d^{(d)} z^d, \tag{4.7}$$

and the coefficients $\frac{1}{k!}P^{(k)}(x)$ of this finite Taylor expansion can be calculated by <u>Horner's algorithm</u>, exactly as in the scalar case (Rall, 1969). As there is much common information in the computation of the value of a polynomial and its derivatives, which makes it possible to design efficient computer programs to perform these operations.

Assuming that $P_d(x+z)=0$ in (4.7) and $[P_d'(x)]^{-1}$ exists, one can obtain the equation

$$z=y+C_2z^2+\ldots C_dz^d \tag{4.8}$$

for $z=x^*-x$, where

$$y= -[P_d(x)]^{-1}P_d(x), \quad C_k= -\frac{1}{k!}[P_d(x)]^{-1}P_d^{(k)}(x), \quad k=2,3,\ldots,d. \tag{4.9}$$

The form (4.8) of equation (4.1) lends itself readily to iterative methods of solution.

The equivalence of general polynomial systems and quadratic systems is based on the following observation.

<u>The polynomial system (4.1) may be replaced by an equivalent quadratic system</u>

$$Bz^2+Az+y = 0 \tag{4.10}$$

<u>which is (i) linear in the coefficients of the original system, and (ii) contains only nonlinear equations of the form</u>

$$z_i-z_jz_k=0, \qquad i>j, \quad i>k, \tag{4.11}$$

<u>where possibly</u> $j=k$.

The only operation required to form products of powers $x_1^{\alpha_1}x_2^{\alpha_2}\ldots{}_n^{\alpha_n}$ of the original variables is multiplication, which is bilinear. One can start with $z_i=x_i$, $i=1,2,\ldots,n$, and then apply (4.11) as often as necessary to obtain a variable z_i corresponding to each distinct term in (4.1). This does not necessarily lead to a unique quadratic system; for example, the scalar quartic equation

$$aw^4+bw^3+cw^2+dw+e = 0 \tag{4.12}$$

is equivalent to the system

$$\left.\begin{aligned} x - wy &= 0, \\ y - wz &= 0, \\ z - w^2 &= 0, \\ ax+by+cz+dw+e &= 0, \end{aligned}\right\} \tag{4.13}$$

or the system

$$\left.\begin{aligned} x - z^2 &= 0, \\ y - wz &= 0, \\ z - w^2 &= 0, \\ ax+by+cz+dw+e &= 0. \end{aligned}\right\} \tag{4.14}$$

The bilinear operator B in (4.10) is extremely <u>sparse</u>, that is, all but a few of its coefficients are zero. One has

$$b_{ijk} = 0, \qquad i \leq n,$$

$$b_{ij_ik_i} = b_{ik_ij_i} = \begin{cases} -\frac{1}{2}, & j_i \neq k_i,\ i > n \\ -1, & j_i = k_i,\ i < n \end{cases} \tag{4.15}$$

where only one value of j_i corresponds to each $i>n$. For the norm (3.1), one has, for example

$$||B|| = 1 \tag{4.16}$$

which simplifies the computation of a scalar majorant.

5. <u>Inversion of power series</u>

At some point x, suppose that the operator P has

the Taylor series expansion (3.28), which converges for $||z||<R$, R positive. If the equation (3.1) has a solution $x+z=x^*$ with $||x^*-x||<R$, then (3.28) becomes

$$-P'(x)z=P(x)+\sum_{k=2}^{\infty}\frac{1}{k!}P^{(k)}(x)z^k. \tag{5.1}$$

If also $[P'(x)]^{-1}$ exists, then (5.1) may be written as

$$z=y+A_2z^2+A_3z^3+\dots, \tag{5.2}$$

where

$$y=-[P'(x)]^{-1}P(x),\quad A_k=-\frac{1}{k!}[P'(x)]^{-1}P^{(k)}(x),\quad k=2,3,4,\dots . \tag{5.3}$$

The form of (5.2) suggests the classical method of _inversion_ (or _reversion_) of a power series for the determination of z. One assumes an expansion

$$z=y+C_2y^2+C_3y^3+\dots \tag{5.4}$$

of z in terms of y, and attempts to determine the multilinear operators $C_2, C_3, \dots$. Formally, this is done by substituting (5.4) into (5.2) and equating the coefficients of the power operators of equal degree. This gives

$$C_2=A_2,\quad C_3=2A_2A_2+A_3,\dots, \tag{5.5}$$

These expressions becoming rapidly more complicated for increasing orders. However, it is not difficult to program a computer to produce the required operators.

In order to justify this procedure, the series

(5.4) must converge to a point z within the radius of convergence R of (5.2). If these conditions are satisfied, then $x^*=x+z$ will be a solution of the original equation (3.1). The most systematic way to verify these results is by the use of scalar majorant series. If

$$r=s+a_2r^2+a_3r^3 +... \qquad (5.6)$$

is a scalar majorant series for (5.2), where $s\geq||y||$, $a_k\geq||A_k||$, $k=2,3,4,...$, then inversion of (5.6) yields the series

$$r=s+c_2s^2+c_3s^3 +... \quad . \qquad (5.7)$$

As $c_k\geq||C_k||$, $k=2,3,4,...$, it follows that the convergence of (5.7) implies the convergence of (5.4), and, if the sum r is within the radius of convergence of (5.6), then the value of z given by (5.4) provides a solution $x^*=x+z$ of the equation (3.1). Writing the series (5.7) as

$$r=s+c_2s^2 +...+ c_ks^k+R_k, \qquad (5.8)$$

with the <u>remainder</u> <u>term</u> R_k, one has for the approximation

$$x^{(k)} = x + z^{(k)} \qquad (5.9)$$

to x^*, where

$$z^{(k)}=y+C_2y^2 +...+ C_kz^k, \qquad (5.10)$$

that

$$||x^*-x^{(k)}||\leq R_k. \qquad (5.11)$$

The analysis of the scalar majorant series thus provides an estimate for the truncation error arising from the use of a finite number of terms of the series (5.4).

The complexity of the method of inversion of power series is reduced somewhat for polynomial systems, as all derivatives of order greater than the degree of the polynomial vanish, and there is no problem concerning the convergence of (4.8). The development of the inverse series is particularly simple for quadratic systems (Rall, 1961, 1969). If

$$z=y+A_2z^2 \tag{5.12}$$

is the system to be solved, then defining

$$C_1y = y \tag{5.13}$$

one has

$$C_2y^2=A_2y^2=A_2(C_1y)^2=A_2(C_1y)(C_1y), \tag{5.14}$$

and, in general,

$$C_ky^k= \sum_{j=1}^{k-1} A_2(C_jy^j)(C_{k-j}y^{k-j}), \tag{5.15}$$

$k=2,3,\ldots$. The scalar majorant series for (5.12) is the polynomial

$$r=s+a_2r^2, \tag{5.16}$$

and the corresponding inverse series (5.7) is simply the binomial expansion of

$$r= \frac{1- \sqrt{1-4a_2s}}{2a_2} , \tag{5.17}$$

which converges provided $a_2s \leq \frac{1}{4}$. In terms of the general quadratic system (4.2), this condition requires that $[Q'(x)]^{-1}$ exists, and

$$||2A_2||\cdot||[Q'(x)]^{-1}||\cdot||-[Q'(x)]^{-1}Q(x)||\leq \frac{1}{2} \qquad (5.18)$$

which, as will be discussed later, is the well-known sufficient condition for the convergence of Newton's iterative method to a solution x^* of (4.2), starting from $x^{(0)}=x$. As the sum (5.17) of the inverse series for (5.16) is known, estimation of the remainder term R_k, and hence the trucation error of a partial sum (5.10) of the series (5.4) is simple.

In view of the large size of systems of equations which often occur in the numerical solution of partial differential equations and the fact that using higher derivatives leads to still larger arrays, it might appear at first glance that the inversion of power series is not an appropriate method. However, it could be practical if the derivatives are sparse and not many terms of the inverse series are required to obtain the desired accuracy of solution. Also, it is not necessary to invert the operator $P'(x)$ to apply this method. One can find y by solving the linear system

$$P'(x)y = - P(x), \qquad (5.19)$$

and C_2y^2 satisfies the linear system,

$$P'(x)(C_2y^2)= - \frac{1}{2} P''(x)y^2, \qquad (5.20)$$

and so on. If the number of terms of the inverse series required is small compared to the order of the system, then this procedure would probably be preferable to the inversion of $P'(x)$. On the other hand, it may be that $[P'(x)]^{-1}$ is easy to obtain for the given choice of x. In this case, the method of inversion of power series may require less effort than an iteration method in which $P'(x)$ has to be updated.

6. Iterative methods

The basic method of iteration consists of the

transformation of the equation (3.1) into an equivalent _fixed point problem_

$$x = F(x) \qquad (6.1)$$

for the _iteration operator_ F. Then, starting from some chosen initial approximation $x^{(0)}$, one generates the sequence $\{x^{(k)}\}$ by use of the relationships

$$x^{(k+1)} = F(x^{(k)}), \quad k=0,1,2,\ldots . \qquad (6.2)$$

By a simple transformation, as applied previously to the polynomial $P_d(x)$ to obtain (4.8), and the power series (3.28) to get (5.2), one could assume that $x^{(0)}=0$, at least for theoretical purposes. However, it will be convenient in what follows to regard $x^{(0)}$ as simply some given point, and look for solutions x^* of (6.1) (and hence of (3.1)) in the _closed ball with center_ $x^{(0)}$ _and radius_ r,

$$\bar{U}(x^{(0)},r)=\{x: \ ||x-x^{(0)}||\leq r\}, \qquad (6.3)$$

or the corresponding _open ball_,

$$U(x^{(0)},r)=\{x: \ ||x-x^{(0)}||< r\} \qquad (6.4)$$

It will also be useful to extend the concept of a _scalar majorant function_ p(r) for an operator P to mean

$$||P(x)|| \leq p(||x-x^{(0)}||) \qquad (6.5)$$

in this context, so that, for example,

$$||P(x)|| \leq p(r) \qquad (6.6)$$

for x in $\bar{U}(x^{(0)},r)$, as majorant functions are monotone increasing. For $r=\infty$, (6.3) and (6.4) represent the entire space X.

The basic result on iteration is the following one, essentially due to Banach (Rall, 1969):

<u>If F satisfies the Lipschitz condition</u>

$$||F(x)-F(z)|| \leq \alpha||x-z||, \qquad \alpha<1, \qquad (6.7)$$

<u>for</u> x,z <u>in</u> $\bar{U}(x^{(0)},r)$, <u>and</u>

$$r \geq \frac{||x^{(0)}-F(x^{(0)})||}{1-\alpha} = r_0, \qquad (6.8)$$

<u>then (6.1) has a unique solution</u> x^* <u>in</u> ($\bar{U}(x^{(0)},r_0)$, <u>and the process (6.2) converges to</u> x^*, <u>with</u>

$$||x^*-x^{(k)}|| \leq \frac{\alpha^k}{1-\alpha}, ||x^{(0)}-F(x^{(0)})||, \quad k=0,1,2,\ldots . \qquad (6.9)$$

As stated above, this so-called <u>contraction mapping principle</u> is a semi-local theorem; it is global if $r=\infty$. The inequalities (6.7) and (6.8) give rise to the possibility of a scalar majorant principle. The Lipschitz constant α in (6.7) will be a monotone increasing function $\alpha(r)$ of the radius of the ball $\bar{U}(x^{(0)},r)$; for example, one may take

$$\alpha(r)=\max||F'(x)||, \quad x \text{ in } \bar{U}(x^{(0)},r), \qquad (6.10)$$

for an operator with a continuous derivative. The satisfaction of the condition (6.8) may then be determined by solving the scalar equation

$$r = \frac{||x^{(0)}-F(x^{(0)})||}{1-\alpha(r)} \qquad (6.11)$$

for positive r. For example, the polynomial equation (4.8) with $x^{(0)}=0$ leads to the scalar equation

$$dc_d r^d+\ldots+2c_2r^2-r+s = 0 \qquad (6.12)$$

where $s=||y||$. By Descartes' rule of signs, (6.12) has two positive solutions (which may coincide) $r_0 \leq r_1$, or none. The value of r_0 gives the radius of the smallest ball in which the existence of x^* can be guaranteed, and

$$\alpha(r_0)=dc_d r^{d-1} +\ldots+2c_2 r \tag{6.13}$$

is the best constant for the error bound (6.9). For quadratic systems, it follows that one must have

$$c_2 s \leq \frac{1}{8}, \tag{6.14}$$

which is more restrictive than the condition $c_2 s \leq \frac{1}{4}$ obtained for the method of inversion of power series. (It will be shown later that simple iteration converges for quadratic equations provided that $c_2 s \leq \frac{1}{4}$.)

Under the hypothesis (6.7), x^* will be unique in $\overline{U}(x^{(0)},r)$ as long as $\alpha(r)<1$. This leads to the equation

$$\alpha(r)=1 \tag{6.15}$$

to determine the largest open ball in which x^* will be unique. For polynomial systems, this becomes

$$dc_d r^{d-1}+\ldots+2c_2 r-1=0 \tag{6.16}$$

which has a unique positive solution R by Descartes' rule of signs. Thus x^* will be unique in $U(x^{(0)},R)$, provided that it exists. For quadratic systems,

$$R = \frac{1}{2c_2} \tag{6.17}$$

Unless α in the error bound (6.9) is extremely small, the method of simple iteration may converge too slowly to be of value in the actual solution of a large system of the type encountered in the numerical solution of a partial differential equation. However, the above theory may be useful in establishing the existence of a solution x^* in a certain region, or in obtaining an error estimate for an approximate solution $x^{(0)}$ computed by some other method. It can also happen that iteration can converge much more rapidly than predicted by (6.9), as the following result shows.

If $x^*=F(x^*)$,

$$F'(x^*)=F''(x^*)=\ldots=F^{(m)}(x^*)=0, \tag{6.18}$$

$F^{(m)}$ *satisfies the Lipschitz condition*

$$||F^{(m)}(x)-F^{(m)}(z)|| \leq (m+1)!K||x-z|| \tag{6.19}$$

for x,z *in* $\overline{U}(x^{(0)},||x^{(0)}-x^*||)$, *and*

$$\theta=K||x^{(0)}-x^*||^m <1 , \tag{6.20}$$

then the sequence $\{x^{(k)}\}$ *defined by (6.2) converges to* x^*, *with*

$$||x^{(k)}-x^*|| \leq \theta^{\frac{(m+1)^k-1}{m}} ||x^{(0)}-x^*||, \quad k=0,1,2,\ldots . \tag{6.21}$$

This local convergence theorem is easy to establish on the basis of Taylor's formula (3.25) and inequality (3.27). For the *Newton iteration operator* N defined by

$$N(x) = x - [P'(x)]^{-1}P(x), \tag{6.22}$$

one has that (6.18) is satisfied with m=1 if $P(x^*)=0$ and $[P'(x^*)]^{-1}$ exists. Thus, one may expect _quadratic_ convergence of $\{x^{(k)}\}$ to x^* if P' is Lipschitz continuous and $x^{(0)}$ is sufficiently close to x^*. As another example, the Čebyšev iteration operator C defined by

$$C(x)=x-[P'(x)]^{-1}P(x)-$$

$$-\frac{1}{2}[P'(x)]^{-1}P''(x)[P'(x)]^{-1}P(x)[P'(x)]^{-1}P(x), \tag{6.23}$$

is such that $C'(x^*)=C''(x^*)=0$, provided that $P(x^*)=0$, $[P'(x^*)]^{-1}$ and $P''(x^*)$ exist, and one may expect _cubic_ convergence (m=2) of $\{x^{(k)}\}$ to x^* if (6.19) and (6.20) are satisfied.

There are many variants and generalizations of the simple iteration process (6.2). For example, there are _multistep_ methods, in which $x^{(k+1)}$ depends a number of previous iterates $x^{(k)},\dots,x^{(k-m+1)}$, and also _nonstationary_ methods, in which the iteration operator takes on the value F_k at the kth step instead of remaining constant, and combinations of these methods. The _secant_ methods, to be discussed later, fall into the category of multistep methods.

An idea adapted from the solution of the linear systems arising from linear partial differential equations is the use of a _relaxation_ (or _over-relaxation_) procedure, in which the iteration makes use of partially updated vectors

$$x^{(m,i)}=(x^{(k+1)},\dots,x_{i-1}^{(k+1)},x_i^{(k)},x_{i+1}^{(k)},\dots,x_m^{(k)}, \tag{6.24}$$

and one solves single equations for $x_i^{(k+1)}$, $i=1,2,\dots,n$, until the entire vector has been updated. As with the transfer of other ideas from linear to nonlinear systems, the implementation of a relaxation method may be hindered by the fact that the equation for $x_i^{(k+1)}$ will in general be nonlinear, and cause difficulties not present in the linear case. However, various methods of this type may be effective in practice (Ortega and Rheinboldt, 1970).

7. Newton's method

In order to be of practical value in the solution of actual systems of nonlinear equations, one would expect that an iterative method should converge quadratically, or nearly quadratically, as the computation of each iterate $x^{(k)}$ may be time-consuming, so that an accurate approximate solution would be desired in as few iterations as possible. At the center of attention from a theoretical and practical standpoint in the development of quadratically convergent iterative processes is Newton's method

$$x^{(k+1)}=x^{(k)}- P (x^{(k)})^{-1}P(x^{(k)}), \quad k=0,1,2,..., \tag{7.1}$$

based on the use of the iteration operator (6.22).

Although (7.1) is written in terms of an inverse operator, in practice it is more efficient to solve the linear systems

$$P'(x^{(k)})y^{(k)}= -P(x^{(k)}), \quad k=0,1,2,..., \tag{7.2}$$

for the differences

$$y^{(k)}=x^{(k+1)}-x^{(k)}, \quad k=0,1,2,..., \tag{7.3}$$

or perhaps

$$P'(x^{(k)})x^{(k+1)}=P'(x^{(k)})x^{(k)}-P(x^{(k)}), \quad k=0,1,2,..., \tag{7.4}$$

for $x^{(k+1)}$ directly. If these systems are large,

an extensive <u>inner</u> iteration process (Young, 1973) may be required for one step of the <u>outer</u> iteration, which is newton's method in this case. The semi-local theorem of Kantorovič (Tapia, 1971) gives sufficient conditions for the convergence of Newton's method.

If $[P'(x^{(0)})]^{-1}$ *exists*,

$$||[P'(x^{(0)})]^{-1}|| \leq \beta, \quad ||-[P'(x^{(0)})]^{-1}P(x^{(0)})|| \leq \eta, \tag{7.5}$$

P' *is Lipschitz continuous with constant* K *in the ball* $\bar{U}(x^{(0)},r)$,

$$h=\beta\eta K \leq \frac{1}{2} \tag{7.6}$$

and

$$r \leq (1-\sqrt{1-2h})\ \frac{\eta}{h} = r^*, \tag{7.7}$$

then the equation $P(x)=0$ *has a solution* x^* *in* $\bar{U}(x^{(0)},r^*)$ *to which the sequence* $\{x^{(k)}\}$ *defined by (7.1) converges*.

The convergence can be shown to be extremely rapid if $h< \frac{1}{2}$ (Ostrowski, 1971; Gragg and Tapia,1973). In fact, for

$$\theta = \frac{1-\sqrt{1-2h}}{1+\sqrt{1-2h}}, \tag{7.8}$$

one has

$$||x^{(k)}-x^*|| \leq \frac{\theta^{2^k}}{1-\theta^{2^k}} \cdot 2\sqrt{1-2h} \cdot \frac{\eta}{h}, \quad k=0,1,2,..., \tag{7.9}$$

which is the best possible error estimate obtainable under the hypotheses given above.

For $h= \frac{1}{2}$, one gets

$$||x^{(k)}-x^*|| \leq (\tfrac{1}{2})^k \ \frac{\eta}{h}, \quad k=0,1,2,\ldots, \tag{7.10}$$

which can also be attained. This rate of convergence is probably too slow to be of practical value, considering the labor involved to obtain the iterates.

The question of the rapid convergence of Newton's method is intimately connected with the invertibility of $P'(x^*)$ (Rall, 1973b). If h $\frac{1}{2}$, then $[P'(x^*)]^{-1}$ exists; on the other hand, if $[P'(x^*)]^{-1}$ exists and P' is Lipschitz continuous, then a Newton sequence $\{x^{(k)}\}$ which converges to x^* will eventually enter a ball in which rapid convergence of the remainder of the sequence can be guaranteed. Thus, it appears that Newton's method will be an effective computationl procedure only for <u>simple</u> solutions x^* of equation (3.1); that is, solutions for which$[P'(x^*)]^{-1}$ exists.

The Kantorovič theorem on the convergence of Newton's method also provides a uniqueness criterion. If $h< \frac{1}{2}$ and

$$r \geq (1+\sqrt{1-2h}) \ \frac{\eta}{h} = r^{**}, \tag{7.11}$$

then x^* is unqiue in the open ball $U(x^{(0)},r^{**})$. (In case $h= \frac{1}{2}$, then $r^*=r^{**}=2\eta$, and the regions of existence and uniqueness is the closed ball $\overline{U}(x^{(0)},2\eta)$.)

Conditions (7.7) and (7.8) may be unified by a single scalar majorant principle, provided that a scalar majorant function $K(r)$ is known for the Lipschitz constant K in the ball $\overline{U}(x^{(0)},r)$ (Rall, 1969).

For

$$h(r) = \beta\eta K(r), \tag{7.12}$$

one looks for solutions of the scalar equation

$$h(r) = \frac{2\eta(r-\eta)}{r^2} \tag{7.13}$$

If (7.13) has positive solutions r^*, r^{**} such that $r^* \leq 2\eta \leq r^{**}$, then these numbers provide the information necessary for the assertions of existence and uniqueness of x^* of the Kantorovič theorem.

For polynomial systems (4.1), one may take

$$K(r)=d(d-1)a_d r^{d-2}+...+2a_2, \tag{7.14}$$

the second derivative of the scalar majorant polynomial for

$$P_d(x)=P_d(x^{(0)})+P_d'(x^{(0)})(x-x^{(0)})+...+\frac{1}{d!}P^{(d)}(x^{(0)})(x-x^{(0)})^d, \tag{7.15}$$

and (7.13) becomes

$$d(d-1)a_d r^d+...+2a_d r^2-\frac{2}{\beta}r+\frac{2}{\beta}\eta = 0, \tag{7.16}$$

which, by Descartes' rule of signs, has two possibly coincident positive solutions r^*, r^{**}. In the case of quadratic systems, $a_2=||A_2||$, and r^*, r^{**} exist provided

$$2a_2\beta\eta \leq \frac{1}{2} \tag{7.17}$$

which is precisely condition (5.18) for the convergence of the inverse power series, as $\beta=||I||=1$ and $\eta=||y||=s$.

Newton's method may be viewed as the use of the first term y of the inverse power series (5.4) at each step as an approximation to $x^*-x^{(k)}$ to obtain $x^{(k+1)}$. The use of more terms of the series gives rise to <u>higher order</u> methods, such as <u>Čebyšev's method</u>, which uses the iteration operator defined by (6.23). It is doubtful if such methods ordinarily offer any computational advantages. For example, to apply Cebysev's method, one must solve the linear systems

Čebyšev's method, one must solve the linear systems

$$P'(x^{(k)})y^{(k)} = -P'(x^{(k)}), P'(x^{(k)})z^{(k)} = -\frac{1}{2}P''(x^{(k)})y^{(k)^2} \quad k=0,1,2,..., \tag{7.18}$$

to obtain

$$x^{(k+1)} = x^{(k)} + y^{(k)} + z^{(k)}, \quad k=0,1,2,... . \tag{7.19}$$

The resulting sequence may converge cubically, but by applying Newton's method twice, one has a fourth-order method which only requires the solution of two linear system, and does not need the value of $P''(x)$.

For the quadratic system (4.2), Newton's method is

$$x^{(k+1)} = (2A_2x^{(k)} + A_1)^{-1}(A_2x^{(k)^2} - A_0), \tag{7.20}$$

or the equivalent linear system. If $A_1=0$, then

$$x^{(k+1)} = \frac{1}{2}(x^{(k)} - [A_2x^{(k)}]^{-1}A_0), \tag{7.21}$$

$k=0,1,2,...$, which is a generalization of the formula known from antiquity for the extraction of square roots.

In some cases, one may wish to consider the application of Newton's method to the fixed point problem (6.1) in order to obtain an iteration process which converges faster than (6.2). The corresponding formula is

$$x^{(k+1)} = x^{(k)} - [I - F'(x^{(k)})]^{-1}[x^{(k)} - F(x^{(k)})], \tag{7.22}$$

where I denotes the <u>identity operator</u> (Ix=x). One may also write (7.22) as

$$x^{(k+1)} = [I - F'(x^{(k)})]^{-1}[F(\alpha^{(k)}) - F'(x^{(k)})x^{(k)}], \tag{7.23}$$

or the corresponding linear systems for k=0,1,2,..., . The following result (Rall, 1972) holds for F uniformly bounded by $\alpha<1$ and Lipschitz continuous in $\bar{U}(x^{(0)},\eta)$, where

$$\eta = \frac{||x^{(0)}-F(x^{(0)})||}{1-\alpha} \tag{7.24}$$

If $||F'(x)||\leq\alpha<1$ *and* $||F'(x)-F'(z)||\leq K||x-z||$ *for* x,z *in* $\bar{U}(x^{(0)},\eta)$, *and*

$$h = \frac{K\,\eta}{1-\alpha} < 2, \tag{7.25}$$

then the sequence $\{x^{(k)}\}$ *defined by (7.22) converges to a fixed point* x^* *of F which is unique in (7.22), with*

$$||x^{(k)}-x^*||\leq(\tfrac{1}{2}h)^{2^{k-1}}\frac{||x(0)-F(x(0))||}{1-\alpha},\quad k=0,1,2,\ldots . \tag{7.26}$$

As one may take $\beta=\frac{1}{1-\alpha}$ in (7.6), this result shows that the assumption of uniform boundedness of F' allows the assertion of quadratic convergence of Newton's method for larger values of h than permitted by the Kantorovič theorem. Scalar majorant principles can also be developed for (7.24) and (7.25).

Supposing that it is not practical to implement Newton's method in an actual case, it is still possible to use the convergence theorems to guarantee the existence of a solution x^*, and estimate the error of an approximate solution $x^{(0)}$ obtained by some other method. Using, for example, the Kantorovič theorem and the methods of interval analysis, these estimates may be made rigorous (Rockne and Lancaster, 1969; Kuba and Rall, 1972).

8. Variants of Newton's method

An iterative procedure of the form

$$x^{(k+1)}=x^{(k)}-A_k^{-1}P(x^{(k)}), \quad k=0,1,2,..., \tag{8.1}$$

in which the linear operators $A_0,A_1,A_2,...$ are considered to be approximations to the derivatives $P'(x^{(0)}, P'(x^{(1)}), P'(x^{(2)}),...$ is called a variant of Newton's method, or a Newton-like method. These methods are frequently employed in case the computation of $P'(x)$ is laborious, or one wishes to have convergence to x^* from an initial approximation $x^{(0)}$ so far away that Newton's method behaves badly. A Newton-like method of the form (8.1) will be said to be consistent if

$$\lim_{k\to\infty} A_k = P'(x^*). \tag{8.2}$$

If the rate of convergence in (8.2) is sufficiently rapid, then the method (8.1) will have a rate of convergence which is quadratic, or nearly so (Ortega and Rheinboldt, 1970). As one does not go to the limit in actual practice, good results may be obtained if A_k is sufficiently close to $P'(x^*)$ or even $P'(x^{(k)})$.

The modified form of Newton's method,

$$x^{(k+1)}=x^{(k)}-[P'(x^{(0)})]^{-1}P(x^{(k)}), \quad k=0,1,2,..., \tag{8.3}$$

is inconsistent unless $P'(x^{(0)})=P'(x^*)$ by chance, in which event it converges quadratically. Ordinarily, (8.3) is slowly convergent, but might be useful because of its simplicity, and it converges under the hypotheses (7.5)-(7.7) of the Kantorovič theorem for Newton's method. It is interesting to note that the modified form of Newton's method and simple iteration for the polynomial equation

$$x=y+C_2x^2+...+C_dx^d \tag{8.4}$$

generate identical sequences starting from $x^{(0)}=0$. This means that for the quadratic system

$$x=y+C_2x^2, \tag{8.5}$$

iteration will converge provided that

$$||y||\cdot||C_2|| \leq \frac{1}{4}, \tag{8.6}$$

which is an improvement over the result (6.14) obtained by application of the contraction mapping principle.

Another method which has been found to be effective in the numerical solution of nonlinear differential equations, but is consistent in only special cases, is the so-called "generalized" Newton method (Greenspan, 1965). It is based on the idea that if the Jacobian matrix $P'(x)$ is diagonally dominant, that is,

$$\left|\frac{\partial p_i}{\partial x_i}\right| > \sum_{\substack{j=1\\ j\neq i}}^{n} \left|\frac{\partial p_i}{\partial x_j}\right|, \quad i=1,2,\ldots,n, \tag{8.7}$$

then $[P'(x)]^{-1}$ exists, and the diagonal matrix

$$[D(x)]^{-1}=\mathrm{diag}\{[\frac{\partial p_1}{\partial x_1}]^{-1}, [\frac{\partial p_2}{\partial x_2}]^{-1},\ldots, [\frac{\partial p_n}{\partial x_n}]^{-1}\} \tag{8.8}$$

may be taken to be an approximation to $[P'(x)]^{-1}$. This gives the iteration process

$$x^{(k+1)}=x^{(k)}-[D(x^{(k)})]^{-1}P(x^{(k)}), \quad k=0,1,2,\ldots, \tag{8.9}$$

with the successive iterates being easy to compute. In another version of this process, one assumes that the system to be solved is of the form

$$Ax + P(x) = 0, \tag{8.10}$$

a linear, and one solves the linear systems

$$[A+D(x^{(k)})]x^{(k+1)}=D(x^{(k)})x^{(k)}-P(x^{(k)}), \tag{8.11}$$

$k=0,1,2,\ldots$, for the successive iterates. The nomenclature "generalized Newton method" is somewhat unfortunate, as this procedure only reduces to the ordinary Newton method if $n=1$, or if $P(x)$ is diagonal. Although one would not ordinarily expect quadratic convergence of this method, it can be of practical utility, particularly if the discrepancy in inequality (8.6) is large.

Another variant of Newton's method which is highly effective in practice is obtained by approximating the partial derivatives in $P'(x)$ by the corresponding <u>difference quotients</u>

$$a_{ij}(x)=\Delta_{j,h}p_i(x)=$$

$$\frac{p_i(x_1,\ldots,x_{j+h},\ldots,x_n)-p_i(x_1,\ldots,x_j,\ldots,x_n)}{h} \tag{8.12}$$

$i,j=1,2,\ldots,n$, and setting $A_k=A(x^{(k)})$ in (8.1). This method will be consistent if $h\to 0$ as $n\to\infty$, provided that the sequence generated converges to x^*. Users of this <u>finite difference form</u> of Newton's method report very good results, particularly if the computation of derivatives requires more time than the corresponding function evaluations. This method thus would appear to be suitable for many types of non-polynomial systems. This procedure can also be combined with elimination (Brown,1967,1969) to obtain a rapidly convergent iteration method, at least for fairly small systems. The finite difference form of Newton's method may also be considered to be a special type of the <u>secant methods</u> discussed in the next section.

9. Secant methods

The basic idea behind Newton's method and the

variants of Newton's method discussed in the previous section may be taken to be the <u>approximation</u> of the nonlinear systems

$$P(x^*-x^{(k)})=0, \quad k=0,1,2,..., \qquad (9.1)$$

by linear systems

$$A_k(x^{(k+1)}-x^{(k)})=y^{(k)}, \quad k=0,1,2,..., \qquad (9.2)$$

to obtain successive iterates. From a geometric point of view, the relationships

$$p_i(x_1,x_2,...,x_n)=y, \quad i=1,2,...,n, \qquad (9.3)$$

may be interpreted as defining <u>surfaces</u> in (n+1)-dimensional space (y is the additional coordinate). In solving equation (3.1), one finds an intersection of these surfaces with the hyperplane y=0 at $(x^*,0)$. Newton's method may be derived by approximating the surfaces (9.3) by their <u>tangent hyperplanes</u> at $x=x^{(k)}$ to obtain the approximating system (9.2) for the intersection of these hyperplanes with y=0. From an intuitive standpoint, this is probably a more satisfactory derivation than regarding Newton's method as the use of the first term of an inverse power series. The variants of Newton's method may also be considered to stem from various linear approximations to the relationships (9.3).

In the case of secant methods, the idea of <u>interpolation</u> comes to the fore. Suppose that one has n+1 points $z^{(0)}, z^{(1)},...,z^{(n)}$ such that the vectors $z^{(j)}-z^{(0)}$, j=1,2,...,n, are linearly independent, as are $P(z^{(j)})-P(z^{(0)})$, j=1,2,...,n. It follows (Ortega and Rheinboldt, 1970) that a matrix A and a vector y exist such that

$$Az^{(j)}+y=P(z^{(j)}), \quad j=0,1,2,...,n. \qquad (9.4)$$

One says that the affine function Ax + y <u>interpolates</u> P(x) at $z^{(0)}, z^{(1)},...,z^{(n)}$. Solving the linear system

$$Ax+y = 0 \tag{9.5}$$

gives the <u>secant approximation</u>

$$x^r = - A^{-1}y$$

to x^*. Under the hypothesis of linear independence given above, x^r exists and is unique (Ortega and Rheinboldt, 1970).

The above method may be converted into an iterative process by associating with an approximation $x^{(k)}$ vectors $z^{(k,0)}, z^{(k,1)},\ldots,z^{(k,n)}$ and performing the interpolation process (9.4) to obtain the linear system

$$A_k x^{(k+1)} + y^{(k)} = 0 \tag{9.6}$$

for $x^{(k+1)}$, $k=0,1,2,\ldots$. The details of these methods vary; however, it is generally true that the iteration process may be carried out by only solving one linear system at each step (Ortega and Rheinboldt, 1970).

A secant method which has proved to be effective in practice is the one due to Robinson (1966). Applied to a system of 24 nonlinear equations obtained by Pack and Swan (1966) in the solution of a magnethohydrodynamic problem, it required the same number of iterations as Newton's method (using exact derivatives) to converge to the same accuracy. Robinson's method required substantially less computer time for this purpose than Newton's method, as it did not require the evaluation of derivatives.

10. <u>Variational and minimization methods</u>

A <u>variational method</u> may be applied to solve the system of equations (3.1) if P is the derivative of a real scalar functional p, that is

$$p\,(x) = P(x)\ . \tag{10.1}$$

The solutions x^* of (3.1) will be <u>critical points</u> of p; conversely, if p has a local minimum (or maximum) at x^*, then x^* will satisfy the system (3.1). A computational method for finding minima or maxima of p is what is called a <u>variational method</u> for solving (3.1). The well-known necessary and sufficient condition for (10.1) to hold if P is differentiable is that the Jacobian matrix $P'(x)$ be symmetric for all x (Vainberg, 1964); that is,

$$\frac{\partial p_i}{\partial x_j} = \frac{\partial p_j}{\partial x_i}, \quad i,j=1,2,\ldots,n. \tag{10.2}$$

This condition is very restrictive in practice, for example, condition (10.2) applied to the polynomial equation (4.1) requires that the coefficient operators $A_1, A_2, \ldots, A_d$ be <u>totally symmetric</u>; that is, for all permutations $q_1\ q_2, \ldots, q_{k+1}$ of the indices $ij_1 \ldots j_k$, one must have

$$a_{ij_1 \ldots j_k} = a_{q_1 q_2 \ldots q_{k+1}} \tag{10.3}$$

It may happen that the partial differential equation being solved numerically has a <u>variational principle</u>, which means that the solution of the partial differential equation provides an extremal value of some real scalar functional. If the approximating system (3.1) is constructed in a manner consistent with the partial differential equation, then one would expect it to have a variational principle also.

Closely related to variational methods are <u>minimization methods</u>, in which one chooses a functional p which has a unique minimum at x=0 (for example, p could be a norm), and minimizes the functional

$$f(x) = p(P(x)). \tag{10.4}$$

If, for example, p is the <u>Euclidean norm</u>

$$||x|| = \left(\sum_{i=1}^{n} x_i^2 \right)^{\frac{1}{2}}, \tag{10.5}$$

then this technique is called a _least-squares_ method for solving (3.1).

There are a number of methods available for the computational minimization of functionals (Daniel, 1971), most of which fall into the category of _descent methods_, which are iterative, and

$$f(x^{(k+1)}) < f(x^{(k)}), \quad k=0,1,2,\ldots \qquad (10.6)$$

for the successive approximations $x^{(0)}, x^{(1)}, x^{(2)}, \ldots$. One of the effective methods in practice is the _Davidon-Fletcher-Powell method_ (Fletcher and Powell, 1963), which is widely used.

It is also possible to develop _hybrid methods_, which combine a straightforward iteration procedure, such as Newton's method, with a descent method to keep the method one of steady reduction of the residual (the value of $p(P(x^{(k)}))$. Powell (1970a, 1970b) has described a method of this type, and published a computer program for its implementation.

It will be unlikely that the computed minimum of f(x) will be exactly zero. The point $x^{(0)}$ at which the minimization method obtains the approximate minimum of f(x) may or may not be close to x^*. In fact, the existence of x^* may not be implied by the fact that f(x) takes on very small values. However, one may apply a theorem of semi-local type in many cases, which will guarantee the existence of x^* and provide an estimate of the error $||x^{(0)}-x^*||$.

11. Continuation methods

Continuation methods operate basically by embedding the problem (3.1) to be solved in a family of problems

$$P(x;\lambda)=0, \qquad 0\leq\lambda\leq 1 \qquad (11.1)$$

in which the solution $x(0)=x^{(0)}$ of

$$P(x;0) = 0 \qquad (11.2)$$

is known, and

$$P(x;1) = P(x), \qquad (11.3)$$

so that $x(1)=x^*$ will be the solution desired for the original system. Continuation methods fall roughly into two categories; discrete, and continuous (or analytic), depending on how the intermediate problems

$$P(x;\lambda)=0, \qquad 0<\lambda<1, \qquad (11.4)$$

are handled.

In discrete continuation (Ficken, 1951) one is usually concerned with the determination of a finite sequence of values of λ,

$$0 = \lambda_1<\lambda_2< \dots <\lambda_{m-1}<\lambda_m=1 \qquad (11.5)$$

for which the solution $x(\lambda_i)$ of the ith problem

$$P(x;\lambda_i) = 0 \qquad (11.6)$$

is a sufficiently good initial approximation to the solution of the (i+1)st problem to insure that some solution method converges to $x(\lambda_{i+1})$, and so on. Thus, for example, if

$$\left.\begin{aligned} &||[P'(x^{(0)};\lambda)]^{-1}|| \le \beta(\lambda), \\ &||-[P'(x^{(0)};\lambda)]^{-1}P(x^{(0)};\lambda)|| \le \eta(\lambda), \\ &||P'(x;\lambda)-P'(z;\lambda)|| \le K(\lambda), \end{aligned}\right\} \qquad (11.7)$$

x,y in $\overline{U}(x^{(0)},r)$, one may determine λ_1 from the scalar inequality

$$h(\lambda)=\beta(\lambda)\eta(\lambda)K(\lambda) \le \frac{1}{2} \qquad (11.8)$$

in order to investigate the applicability of Newton's method to the first intermediate problem. In actual practice, one would be faced with the decision of choosing λ_{i+1} so that $h(\lambda_{i+1})$ would be small enough to insure that the intermediate problem for $x(\lambda_{i+1})$ is solved with sufficient accuracy, and yet keep $\lambda_{i+1}-\lambda_i$ large enough so that the number of intermediate problems to be solved is not too great. If it is not possible to bound $\lambda_{i+1}-\lambda_i$ uniformly from below, then a discrete continuation method cannot be guaranteed to exist for the given problem.

If the only goal of a continuation method is to obtain $x^*=x(1)$, then the intermediate problems do not have to be solved in all cases with great accuracy in order to keep advancing upon x^*. On the other hand, the parameter λ may have some useful significance, and the table of solutions of the intermediate problems would need to be computed as accurately as the final result.

Another type of discrete continuation method which may occur in the numerical solution of partial differential equations is what might be called dimensional continuation. An accurate numerical solution may require the solution of a very large system, which in turn would need an accurate initial approximation to its solution-vector in order to be solvable in a reasonable amount of time. A less accurate approximation to the solution of the differential equation would give a system of smaller size, but of the same structure as the larger system. Interpreting the solutions of the smaller system as approximate solutions of the partial differential equation (for example, as values of the solution on a coarser mesh than known to be needed) may allow an interpolation process to give a sufficiently good result to start the solution of the large system. This process of solving larger and larger approximating systems to the partial differential equation may also be considered in the framework of continuation, with the parameter λ dependent on the size of the system.

As an example of dimensional continuation, the finite difference approximations to the Dirichlet problem on a square for

$$\Delta u = u^2, \tag{11.9}$$

with homogeneous boundary conditions, have been solved for their negative solutions. Once these have been obtained for a relatively large mesh size (Rall, 1969), linear interpolation of their values gives useful initial approximations for the equations on a refined mesh.

Analytic continuation methods attempt to determine the solution curve

$$x=x(\lambda), \qquad 0\le\lambda\le 1 \tag{11.10}$$

of (11.1). One method of this kind is due to Davidenko (1953). A differentiable arc $h(\lambda)$ is chosen such that

$$h(0) = P(x^{(0)}), \ h(1) = 0, \tag{11.11}$$

for example,

$$h(\lambda) = (1-\lambda)P(x^{(0)}). \tag{11.12}$$

The solution curve $x(\lambda)$ is defined implicitly by

$$P(x) = h(\lambda). \tag{11.13}$$

Differentiating (11.13) with respect to λ, setting $h'(\lambda)=g(\lambda)$, gives

$$P'(x)\frac{dx}{d\lambda} = g(\lambda), \tag{11.14}$$

and thus one has the initial value problem

$$\left.\begin{aligned} \frac{dx}{d\lambda} &= [P'(x)]^{-1}g(\lambda), \\ x(0) &= x^{(0)} \end{aligned}\right) \tag{11.15}$$

to integrate over $0\le\lambda\le 1$ to obtain $x(\lambda)$ and thus $x^*=x(1)$. A numerical integration of (11.15) gives an approximate solution of (3.1).

While the integration of initial value problems, even fairly large ones, can be done by effective computer programs, the evaluation of the right hand side of (11.15) does require repeated solution of the linear system (11.14). As

$$\Gamma=[P'(x)]^{-1} \tag{11.16}$$

satisfies the differential equation

$$\frac{d\Gamma}{d\lambda} = -\Gamma\, P''(x)\, \frac{dx}{d\lambda}\, \Gamma\ , \tag{11.17}$$

(Rall, 1968), one may obtain the larger initial value problem

$$\left.\begin{aligned} \frac{dx}{d\lambda} &= \Gamma\, g(\lambda), \quad x(0)=x^{(0)}, \\ \frac{d\Gamma}{d\lambda} &= -\Gamma\, P''(x)\Gamma g(\lambda)\Gamma, \quad \Gamma(0)=[P'(x^{(0)})]^{-1}, \end{aligned}\right) \tag{11.18}$$

which requires only the computation of $[P'(x^{(0)})]^{-1}$, and might be possible to use if $P''(x)$ is sparse. If P is a quadratic polynomial operator, then $P''(x)=2A_2$, a constant, and (11.18) becomes

$$\left.\begin{aligned} \frac{dx}{d\lambda} &= \Gamma g(\lambda) \quad , \quad x(0)=x^{(0)}, \\ \frac{d\Gamma}{d\lambda} &= -2\Gamma A_2\Gamma g(\lambda)\Gamma, \ \Gamma(0)=[P'(x^{(0)})]^{-1} \end{aligned}\right) \tag{11.19}$$

The second system does not contain x, and can be solved independently of the first. Once $\Gamma(\lambda)$ is known, $x(\lambda)$ is obtained from the first equation by simple integration. Integration of (11.19) for scalar quadratic equations yields the usual solution formula.

Continuation methods work, at least in theory, if x^* exists and the solution curve $x(\lambda)$ is sufficiently well behaved, and do not depend on $x^{(0)}$ being close to x^*. The solution of many intermediate problems or the integration of large initial value problems can be laborious, so that a continuation method may well be a last resort. However, one has the advantage of having the implicit expression (11.1) (or (11.13) in the case of Davidenko's method) for the solution curve $x(\lambda)$, if it exists.

12. Acknowledgements

This work was supported by the Science Research Council of Great Britain, and the U.S. Army under contract No. DA-31-124-ARO-D-462. Particular thanks is due to Professor Leslie Fox of the Oxford University Computing Laboratory and Professor L.M. Delves of the Department of Computational and Statistical Science, University of Liverpool, for the use of the facilities of their departments.

REFERENCES

BAKER, C.T.H., DELVES, L.M., AND WALSH, J. (1973) (EDS.) NUMERICAL SOLUTION OF INTEGRAL EQUATIONS, OXFORD (TO APPEAR).

BROWN, K. (1967) Solution of simultaneous nonlinear equations, Comm. ACM 10, 728-729.

BROWN, K. (1969) A quadratically convergent Newton-like method based on Gaussian elimination, SIAM J. Numer. Anal. 6, 560-569.

DANIEL, J.W. (1971) THE APPROXIMATE MINIMIZATION OF FUNCTIONALS, PRENTICE-HALL.

DAVIDENKO, D.F. (1953) On a new method of numerically integrating a system of nonlinear equations (Russian), Dokl. Akad. Nauk. SSSR 88, 601-604.

DEJON, B. AND HENRICI, P. (1969) (EDS.) CONSTRUCTIVE ASPECTS OF THE FUNDAMENTAL THEOREM OF ALGEBRA, WILEY-INTERSCIENCE.

DIEUDONNÉ, J. (1960) FOUNDATIONS OF MODERN ANALYSIS, ACADEMIC PRESS.

DUFFING, G. (1918) ERZWINGENE SCHWINGUNGEN BEI VERÄNDERLICHER EIGENFREQUENZ UND IHRE TECHNISCHE BEDEUTUNG, VIEWEG & SOHN.

DURAND, E. (1960) SOLUTIONS NUMERIQUES DES EQUATIONS ALGEBRIQUES. TOM I: ÉQUATIONS DU TYPE F(x)=0; RACINES D'UN POLYNOME, MASSON ET CIE.

FLETCHER, R. and POWELL, M.J.D. (1963) A rapdly convergent descent method for minimization, Comput. J. 6, 163-168.

GRAGG, W.B. and TAPIA, R.A. (1973) Optimal error bounds for the Newton-Kantorovich theorem, SIAM J. Numer. Anal. (to appear).

GREENSPAN, D. (1965) INTRODUCTORY NUMERICAL ANALYSIS OF ELLIPTIC BOUNDARY VALUE PROBLEMS, HARPER AND ROW.

HANSEN, E. (1969) (ED.) TOPICS IN INTERVAL ANALYSIS OXFORD.

HOUSEHOLDER, A.S. (1970) THE NUMERICAL TREATMENT OF A SINGLE NONLINEAR EQUATION, MCGRAW-HILL.

KRASNOSEL'SKII, M.A., VAINIKKO, G.M., ZABREIKO, P.P., RUTITSKII, YA.B., AND STETSENKO, V.YA. (1972) APPROXIMATE SOLUTION OF OPERATOR EQUATIONS, TR. BY D. LOUVISH, WOLTERS-NOORDHOFF.

KUBA, D. AND RALL, L.B. (1972) A UNIVAC 1108 PROGRAM FOR OBTAINING RIGOROUS ERROR ESTIMATES FOR APPROXIMATE SOLUTIONS OF SYSTEMS OF EQUATIONS, TECH. SUMMARY RPT. NO. 1168, MATH. RESEARCH CENTER, UNIV. OF WISCONSIN.

MACDUFFEE, C.C. (1954) THEORY OF EQUATIONS, WILEY.

MOORE, R.E. (1966) INTERVAL ANALYSIS, PRENTICE-HALL.

ORTEGA, J.M. AND RHEINBOLDT, W.C. (1970) ITERATIVE SOLUTION OF NONLINEAR EQUATIONS IN SEVERAL VARIABLES, ACADEMIC PRESS.

OSTROWSKI, A.M. (1966) SOLUTION OF EQUATIONS AND SYSTEMS OF EQUATIONS (2ND ED.), ACADEMIC PRESS.

OSTROWSKI, A.M. (1970) La Méthode de Newton dans les espaces de Banach, C.R. Acad. Sci. Paris Sér. A 272,1251-1253.

PACK, D.C. and SWAN, G.W. (1966) Magnetogasdynamic flow over a wedge, J. Fluid Mech. 25, 165-178.

POWELL, M.J.D. (1970a) A hybrid method for nonlinear equations, Rabinowitz (1970), 87-114.

POWELL, M.J.D. (1970b) A FORTRAN subroutine for solving systems of nonlinear equations, Rabinowitz (1970), 115-161.

RABINOWITZ, P. (1970) (ED.) NUMERICAL METHODS FOR NONLINEAR ALGEBRAIC EQUATIONS, GORDON AND BREACH.

RALL, L.B. (1961) Quadratic equations in Banach spaces, Rend. Circ. Mat. Palermo 10, 314-332.

RALL, L.B. (1965) (ED.) ERROR IN DIGITAL COMPUTATION (2 VOLS.), WILEY.

RALL, L.B. (1968) DAVIDENKO'S METHOD FOR THE SOLUTION OF NONLINEAR OPERATOR EQUATIONS, TECH. SUMMARY RPT. NO. 948, MATH. RESEARCH CENTER, UNIV. OF WISCONSIN.

RALL, L.B. (1969) COMPUTATIONAL SOLUTION OF NONLINEAR OPERATOR EQUATIONS, WILEY.

RALL, L.B. (1972) CONVERGENCE OF STIRLING'S METHOD IN BANACH SPACES, TECH. SUMMARY RPT. NO. 1249, MATH. RESEARCH CENTER, UNIV OF WISCONSIN.

RALL, L.B. (1973a) Theory of nonlinear integral equations, Baker et al (1973) (to appear).

RALL, L.B. (1973b) A note on the convergence of Newton's method, SIAM J. Numer. Anal. (to appear).

ROKNE, J. and LANCASTER, P. (1969) Automatic errorbounds for the approximate solution of equations, Computing 4, 294-303.

ROBINSON, S.M. (1966) Interpolative solution of systems of nonlinear equations, SIAM J. Numer. Anal. 3, 650-658.

TAPIA, R.A. (1971) The Kantorovich theorem for Newton's method, Amer. Math. Monthly 78, 389-392.

TAYLOR, A.E. (1958) INTRODUCTION TO FUNCTIONAL ANALYSIS, WILEY.

TRAUB, J.F. (1964) ITERATIVE METHODS FOR THE SOLUTION OF EQUATIONS, PRENTICE-HALL.

VAINBERG, M.M. (1964) VARIATIONAL METHODS FOR THE STUDY OF NONLINEAR OPERATORS, TR. BY A. FEINSTEIN, HOLDEN-DAY.

WILKINSON, J.H. (1963) ROUNDING ERRORS IN ALGEBRAIC PROCESSES, PRENTICE-HALL.

WILKINSON, J.H. (1965) THE ALGEBRAIC EIGENVALUE PROBLEM, OXFORD.

YOUNG, D.M. (1973) Solution of linear systems of equations, This Volume.

ZAGUSKIN, V.L. (1961) HANDBOOK OF NUMERICAL METHODS FOR THE SOLUTION OF ALGEBRAIC EQUATIONS, TR. BY G.O. HARDING, PERGAMON.

ELEMENT TYPES AND BASE FUNCTIONS

A.R. Mitchell

University of Dundee, U.K.

1. THE TRIANGLE

Lagrangian interpolation

The triangle or two dimensional simples is probably the most widely used finite element. One reason for this is that arbitrary regions in two dimensions can be approximated by polygons, which can always be divided up into a finite number of triangles. In addition, the complete m th order polynomial

$$\Pi_m(x,y) = \sum_{i+j=0}^{m} \alpha_{ij} x^i y^j \qquad (1)$$

can be used to interpolate a function, say $U(x,y)$, at $\frac{1}{2}(m+1)(m+2)$ symmetrically placed nodes in a triangle. The first three cases of this general representation for the triangle $P_1P_2P_3$, with the coordinates of the vertices being (x_1,y_1), (x_2,y_2) and (x_3,y_3) respectively, are

(i) the linear case $(m=1)$. Here the polynomial is

$$\Pi_1(x,y) = \alpha_1 + \alpha_2 x + \alpha_3 y = \sum_{r=1}^{3} U_r p_r^{(1)}(x,y)$$

where $U_r(r=1,2,3)$ are the values of $U(x,y)$ at the vertices

P_1, P_2, P_3 respectively

$$p_r^{(1)}(x,y) = \tau_r + \eta_r x - \xi_r y)/|A|, \qquad (2)$$

with

$$\tau_1 = x_2 y_3 - x_3 y_2$$
$$\xi_1 = x_2 - x_3$$
$$\eta_1 = y_4 - y_3,$$

and τ_r, ξ_r, η_r defined cyclically for $r = 2,3$. The determinant $|A|$, whose modulus is twice the area of the triangle $P_1 P_2 P_3$ is non-zero if P_1, P_2 and P_3 are not collinear. It is easily seen that

$$p_r(x_s, y_s) = \begin{matrix} 1 & (s = r) \\ 0 & (s \neq r) \end{matrix}$$

where $r, s = 1,2,3$.

(ii) the quadratic case ($m=2$). The polynomial is now

$$\Pi_2(x,y) = \beta_1 + \beta_2 x + \beta_3 y + \beta_4 x^2 + \beta_5 xy + \beta_6 y^2$$
$$= \sum_{r=1}^{6} U_r p_r^{(2)}(x,y), \qquad (3)$$

where $U_r (r = 1,2,3)$ are the values of $U(x,y)$ at the vertices P_1, P_2, P_3 and $U_r (r=4,5,6)$ are the values of $U(x,y)$ at P_4, P_5, P_6 the mid points of sides $P_1 P_2$, $P_2 P_3$, $P_3 P_1$ respectively. The functions $p_r^{(2)}(x,y)$ $(r = 1,2,\ldots,6)$ are given by

$$p_1^{(2)}(x,y) = p_1^{(1)}(2p_1^{(1)} - 1),$$

with $p_2^{(2)}(x,y)$ and $p_3^{(2)}(x,y)$ similarly, and

$$p_4^{(2)}(x,y) = 4p_1^{(1)}p_2^{(1)}$$

with $p_5^{(2)}(x,y)$ and $p_6^{(2)}(x,y)$ similarly. Again it follows that

$$p_r^{(2)}(x_s,y_s) = \begin{matrix} 1 & (s = r) \\ 0 & (s \neq r) \end{matrix}$$

where $r,s = 1,2,\ldots,6$. This is a particularly satisfactory result: the basis functions $p_r^{(2)}(x,y)$ $(r = 1,2,\ldots,6)$ for piecewise quadratic approximation over a triangular network can be expressed as quadratic expressions involving only the basis functions $p_r^{(1)}(x,y)$ $(r = 1,2,3)$ for the linear case.

(iii) the cubic case (m=3). The polynomial is

$$\Pi_3(x,y) = \gamma_1+\gamma_2x+\gamma_3y+\gamma_4x^2+\gamma_5xy+\gamma_6y^2+\gamma_7x^3+\gamma_8x^2y+\gamma_9xy^2+\gamma_{10}y^3$$
$$= \sum_{r=1}^{10} U_r p_r^{(3)}(x,y) \qquad (4)$$

where $U_r (r = 1,2,3)$ are the values at the points of trisection of the sides, and U_{10} is the value of $U(x,y)$ at the centroid of the triangle. The basis functions are given by

$$p_1^{(3)}(x,y) = \tfrac{1}{2}p_1^{(1)}, (3p_1^{(1)} - 1)(3p_1^{(1)} - 2),$$

with $p_2^{(3)}(x,y)$ and $p_3^{(3)}(x,y)$ similarly,

$$p_4^{(3)}(x,y) = 9/2\; p_1^{(1)}p_2^{(1)}(3p_1^{(1)} - 1)$$

$$p_5^{(3)}(x,y) = 9/2\; p_1^{(1)}p_2^{(1)}(3p_2^{(1)} - 1)$$

with $p_6^{(3)},\ldots,p_9^{(3)}$ similarly, and

$$p_{10}^{(3)}(x,y) = 27p_1^{(1)}p_2^{(1)}p_3^{(1)}.$$

The tenth parameter is often eliminated by using the linear relation

$$U_{10} = \tfrac{1}{4}(U_4+U_5+U_6+U_7+U_8+U_9) - \tfrac{1}{6}(U_1+U_2+U_3)$$

For a discussion of the elimination of internal points in a general simplex, see Ciarlet and Raviart (3).

We now turn to the general case where the complete m th order polynomial is given by (1). This polynomial has $\frac{1}{2}(m+1)(m+2)$ coefficients which can be chosen so that the polynomial interpolates $U(x,y)$ at the $\frac{1}{2}(m+1)(m+2)$ symmetrically placed points on the triangle $P_1P_2P_3$ whose coordinates are given by

$$\left(\sum_{t=1}^{3} \beta_t x_t/m, \sum_{t=1}^{3} \beta_t y_t/m\right), \qquad (5)$$

where β_1, β_2, and β_3 are integers satisfying $0 \leq \beta_t \leq m$, $(t = 1,2,3)$, and $\beta_1 + \beta_2 + \beta_3 = m$. These points include the three vertices of the triangle $P_1P_2P_3$. The remaining points are obtained geometrically by dividing each side of the triangle into m equal parts and joining the points of subdivision by lines parallel to the sides of the triangle. This subdivides the triangle into m^2 congruent triangles whose vertices are the $\frac{1}{2}(m+1)(m+2)$ points described by (5.). If U_r denotes the value of $U(x,y)$ at a point given by (5), the interpolating polynomial of degree m can be expressed as

$$U(x,y) = \sum_{r=1}^{\frac{1}{2}(m+1)(m+2)} U_r p_r^{(m)}(x,y) \qquad (6)$$

where the summation is over all $\frac{1}{2}(m+1)(m+2)$ points, and $p_r^{(m)}(x,y)$ is a polynomial basis function of degree m taking the value unity at the point associated with the triple $(\beta_1,\beta_2,\beta_3)$ and the values zero at every other point. The formula (6) is a Lagrangian type interpolating formula.

The Standard Triangle

It should be noted from (2) that

(i) $$\sum_{r=1}^{3} p_r^{(1)}(x,y) = 1,$$

(ii) the linear equations $p_r^{(1)}(x,y) = 0 (r = 1,2,3)$ represent the triangle sides $P_2 P_3, P_3P_1$ and P_1P_2 respectively.

(iii) $p_r^{(1)}(x,y) = 1 (r = 1,2,3)$ at the vertices P_1, P_2 and P_3 respectively. Alternatively, the triangle $P_1P_2P_3$ in the (x,y) plane is transformed into the standard triangle $\Pi_1\Pi_2\Pi_3$ in the $(p_1^{(1)}, p_2^{(1)})$ plane by the transformation formulae (2), where $\Pi_1 \equiv (1,0)$, $\Pi_2 \equiv (0,1)$ and $\Pi_3 \equiv (0,0)$. The reverse transformation from the $(p_1^{(1)}, p_2^{(1)})$ plane to the (x,y) plane is given by

$$\begin{aligned} x &= x_3 + (x_1-x_3)p_1^{(1)} + (x_2-x_3)p_2^{(1)} \\ y &= y_3 + (y_1-y_3)p_1^{(1)} + (y_2-y_3)p_2^{(1)} \end{aligned} \qquad (7)$$

Since all triangles in a triangular network in the (x,y) plane can be transformed into this standard triangle, it is very convenient to work in terms of this standard triangle, and at an appropriate point to transfer the result bank to a particular triangle in the (x,y) plane through the linear transformation (7). This procedure will be used repeatedly when triangular elements are involved.

C^0 Approximating Functions

In the general case, the function interpolates the value of $U(x,y)$ at $\frac{1}{2}(m+1)(m+2)$ points on the triangle. On a side of the triangle, this function reduces to a polynomial of degree m in the variable s which is measured along the side of the triangle. The polynomial interpolates $U(x,y)$ at $(m+1)$ points on the side of the triangle, and so is unique. Also in a triangular network, each side (internal) belongs to two triangles. If the function in each triangle interpolate at $\frac{1}{2}(m+1)(m+2)$ symmetrically placed points, they will reduce to the unique polynomial of degree m in s on the common side. This means that the interpolating function over the complete triangular network is continuous along internal sides of the network and so has C^0 continuity over the polynomial region.

Hermitian Interpolation

As an alternative to interpolating the function $U(x,y)$ at a large number of points symmetrically placed on the triangle, it is possible to interpolate $U(x,y)$ and some of its derivatives at a smaller number of points.

A class of polynomials suited to this particular task consists of the complete polynomials $p_\nu(x,y)$ of odd degree $2\nu+1$ $(\nu = 1,2,3,\ldots)$, which are determined by the values

$$D^i p_\nu(p_j), \qquad |i| \leq \nu \qquad j = 1,2,3$$

$$D^i p_\nu(p_4), \qquad |i| \leq \nu - 1.$$

Here P_1, P_2, P_3 are the vertices of the triangle, and P_4 is the centroid; $i = (i_1, i_2)$, where i_1, i_2 are nonnegative integers,

$|i| = i_1 + i_2$ and $D^i p = \frac{\partial^{|i|} p}{\partial x^{i_1} \partial y^{i_2}}$. The first two cases of this general representation are

(i) the cubic case ($\nu = 1$). Here the complete cubic polynomial has ten coefficients which are uniquely determined by matching the function values and the first order partial derivatives at the vertices and the function value at the centroid. Thus in this case we may write the polynomial $\Pi_3(x,y)$ which is given in (4), in the form

$$\Pi_3(x,y) = \sum_{r=1}^{4} U_r q_r^{(3)}(x,y) + \sum_{r=1}^{3} [(\frac{\partial U}{\partial x})_r \, r_r^{(3)}(x,y) + (\frac{\partial U}{\partial y})_r \, s_r^{(3)}(x,y)] \tag{9}$$

where

$$q_1^{(3)}(x,y) = p_1^{(1)}(3p_1^{(1)} - 2p_1^{(1)^2} - 7p_2^{(1)} p_3^{(1)})$$

with $q_2^{(3)}(x,y)$, and $q_3^{(3)}(x,y)$ cyclically. At the centroid

$$q_4^{(3)}(x,y) = 27 p_1^{(1)} p_2^{(1)} p_3^{(1)}.$$

Also

$$r_1^{(3)}(x,y) = p_1^{(1)} [\xi_3 p_2^{(1)} (p_3^{(1)} - p_1^{(1)}) + \xi_2 p_3^{(1)} (p_1^{(1)} - p_2^{(1)})]$$

and $s_1^{(3)}(x,y)$ is obtained from $r_1^{(3)}(x,y)$ by replacing ξ by η. The remaining $r_r^{(3)}(x,y)$ and $s_r^{(3)}(x,y)$ are obtained cyclically. This element is in common use in the finite element method.

(ii) the quintic case ($\nu = 2$). This time the complete quintic polynomial has twenty-one coefficients which are uniquely determined by matching the function values and the first and second order partial derivatives at the vertices, and the function value and the first order partial derivatives at the centroid. This element is of little practical use and will not be considered further.

It is worth mentioning a device suggested by Zlamal for eliminating the tenth parameter in the cubic case ($\nu = 1$). The parameter U_4 (value of U at the centroid) is replaced by a suitable linear combination of the other nine parameters U_i, $\frac{\partial U}{\partial x_i}$, $\frac{\partial U}{\partial y_i}$ $(i = 1,2,3)$. The form suggested by Zlamal was

$$U_4 = \tfrac{1}{3}(U_1+U_2+U_3) + \tfrac{1}{18}\,[(x_2+x_3-2x_1)(\tfrac{\partial U}{\partial x})_1$$

$$+ (y_2+y_3-2y_1)(\tfrac{\partial U}{\partial y})_1 + (x_1+x_3-2x_2)(\tfrac{\partial U}{\partial x})_2$$

$$+ (y_1+y_3-2y_2)(\tfrac{\partial U}{\partial y})_2 + (x_1+x_2\,2x_3)(\tfrac{\partial U}{\partial x})_3$$

$$+ (y_1+y_2-2y_3)(\tfrac{\partial U}{\partial y})_3],$$

and for the derivation of this form, the reader is referred to Zlamal (16). If this value of U_4 is substituted into (9), the polynomial $\Pi_3(x,y)$, which is now written as $\Pi_3^*(x,y)$, becomes

$$\Pi_3^*(x,y) = \sum_{r=1}^{3} U_r q_r^{(3)*} + \sum_{r=1}^{3} [(\tfrac{\partial U}{\partial x})_r\, r_r^{(3)*} + (\tfrac{\partial U}{\partial y})_r\, s_r^{(3)*}] \qquad (10)$$

where

$$q_1^{(3)*}(x,y) = p_1^{(1)}(3p^{(1)} - 2p^{(1)^2} + 2p_2^{(1)}p_3^{(1)}) \qquad (10a)$$

with $q_2^{(3)*}(x,y)$ and $q_3^{(3)*}(x,y)$ cyclically. Also

$$r_1^{(3)*}(x,y) = p_1^{(1)^2}(p_3^{(1)}\xi_2 - p_2^{(1)}\xi_3) + \tfrac{1}{2}p_1^{(1)}p_2^{(1)}p_3^{(1)}(\xi_2-\xi_3) \tag{10b}$$

and the formula for $s_1^{(3)*}(x,y)$, referred to as (10c), is obtained from $r_1^{(3)*}(x,y)$ by replacing ξ by η. The remaining $r_r^{(3)*}(x,y)$ and $s_r^{(3)*}(x,y)$ are obtained cyclically.

Tricubic Interpolation

Birkhoff (1) introduced a triangular element which involves the 12 parameter family of all quartic polynomials which are cubic along any parallel to any side of a triangle. With respect to the standard triangle, such a family is

$$\begin{aligned} U(p,q) = {} & a_1+a_2p+a_3q+a_4p^2+a_5pq+a_6q^2+a_7p^3+a_8p^2q \\ & +a_9pq^2+a_{10}q^3+a_{11}(p+q)p^2q+a_{12}(p+q)pq^2 \end{aligned} \tag{11}$$

where for convenience, we are using p and q instead of $p_1^{(1)}$ and $p_2^{(1)}$. The polynomial (11) is called a tricubic polynomial. This polynomial uniquely interpolates given values of U, $\frac{\partial U}{\partial p}$, $\frac{\partial U}{\partial q}$ and $\frac{\partial^2 U}{\partial r \partial s}$ at each vertex, where $\frac{\partial^2 U}{\partial r \partial s}$ is a cross derivative determined at each vertex from the derivatives $\frac{\partial}{\partial r}$ and $\frac{\partial}{\partial s}$ parallel to the adjacent sides. The unique interpolating function takes the form

$$\begin{aligned} U(p,q) = \sum_{i=1}^{3} \Big[& U_i\alpha_i(p,q) + \left(\frac{\partial U}{\partial p}\right)_i \beta_i(p,q) + \left(\frac{\partial U}{\partial q}\right)_i \gamma_i(p,q) \\ & + \left(\frac{\partial^2 U}{\partial r \partial s}\right)_i \delta_i(p,q)\Big], \end{aligned} \tag{12}$$

where the suffix i denotes the value of the quantity at the vertex Π_i of the standard triangle. The coefficients are given by

$$\alpha_1 = p^2(3-2p+6qr) \qquad (\alpha_2,\alpha_3 \text{ cyclically})$$
$$\beta_1 = p^2(p-1-4qr)$$
$$\beta_2 = pq^2\ (1+2r)$$
$$\beta_3 = pr^2(1+2q)$$
$$\gamma_1 = p^2q(1+2r),$$
$$\gamma_2 = q^2(q-1-4pr)$$
$$\gamma_3 = qr^2(1+2p)$$
$$\delta_1 = 2p^2qr$$
$$\delta_2 = 2pq^2r$$
$$\delta_3 = 2pqr^2$$

where $r = 1 - p - q$. The cross derivatives $\left(\frac{\partial^2 U}{\partial r \partial s}\right)_i$ $(i = 1,2,3)$ are given by $-\frac{\partial^2 U}{\partial p \partial Q}$, $\frac{\partial^2 U}{\partial q \partial Q}$ and $\frac{\partial^2 U}{\partial p \partial q}$ respectively where $Q = q - p$. The unique tricubic interpolating polynomial in a particular triangle in the (x,y) plane is obtained from (12) by using the linear transformation formulae (2).

C' Approximating Functions

A triangular element is now introduced which involved the complete family of quintic polynomials. In the plane of the standard triangle, the complete quintic is

$$U(p,q) = a_1+a_2p+a_3q+\ldots,+a_{16}p^5+a_{17}p^4q+a_{18}p^3q^2 +a_{19}p^2q^3+a_{20}pq^4+a_{21}q^5 \qquad (13)$$

and the coefficients a_i $(i = 1,\ldots,21)$ can be uniquely determined in terms of the 21 parameters

$$U, \frac{\partial U}{\partial p}, \frac{\partial U}{\partial q}, \frac{\partial^2 U}{\partial p^2}, \frac{\partial^2 U}{\partial p \partial q}, \frac{\partial^2 U}{\partial q^2} \qquad \text{at each vertex}$$

$$\frac{\partial U}{\partial n} \qquad \text{at each mid side point,}$$

where n is in the direction of the perpendicular to the side. The function U(p,q) given by (13) reduces to a quintic in s along each side of the triangle, where s is p,q-p and q respectively. Each quintic in s is uniquely determined by the parameters at the vertices which provide six boundary conditions, viz. U, $\frac{\partial U}{\partial s}$, $\frac{\partial^2 U}{\partial s^2}$ at the end points of each side. The normal derivative to each side, $\frac{\partial U}{\partial n}$, where n is q,p+q, and p respectively is a quartic in s, and is uniquely determined by the parameters $\frac{\partial U}{\partial n}$, $\frac{\partial^2 U}{\partial n \partial s}$ at the end points of each side, together with $\frac{\partial U}{\partial n}$ at the mid side point. It has thus been shown that the complete quintic polynomials produce an approximating function over a network of triangular elements, which has continuity of displacement and gradient over the complete region. Such an interpolating function is said to be C' over the region.

In fact the parameters corresponding to the normal derivatives at the mid side points can be eliminated without destroying the C' continuity over the triangular network. This is accomplished by imposing a cubic variation of the normal derivative along each side, which is equivalent to putting

$$a_{17} = a_{20} = 0$$

$$5a_{16}+a_{18}+a_{19}+5a_{21}=0$$

in (13). This reduces (13) to an eighteen parameter family of polynomials, and the unique interpolating function is given by

$$U(p,q) = \sum_{i=1}^{3} [U_i \alpha_i(p,q) + (\frac{\partial U}{\partial p})_i \beta_i(p,q) + (\frac{\partial U}{\partial q})_i \gamma_i(p,q) \tag{14a}$$

$$+ (\frac{\partial^2 U}{\partial p \partial q})_i \delta_i(p,q) + (\frac{\partial^2 U}{\partial p^2})_i \epsilon_i(p,q) + (\frac{\partial^2 U}{\partial q^2})_i \Phi_i(p,q)]$$

where the coefficients are given by

$$\begin{aligned}
\alpha_1 &= p^2(10p-15p^2+6p^3+15q^2r)\\
\alpha_2 &= q^2(10q-15q^2+6q^3+15p^2r)\\
\alpha_3 &= r^2(10r-15r^2+6r^3+30pq(p+q))\\
\beta_1 &= p^2(-4p+7p^2-3p^3-15/2\ q^2r)\\
\beta_2 &= pq^2(3-2q-3/2\ p-3/2\ p^2+3/2\ pq)\\
\beta_3 &= pr^2(3-2r-3p^2+6pq)\\
\gamma_1 &= p^2q(3-2p-3/2\ q-3/2\ q^2+3/2\ pq)\\
\gamma_2 &= q^2(-4q+7q^2-3q^3-15/2\ p^2r)\\
\gamma_3 &= qr^2(3-2r-3q^2+6pq)\\
\delta_1 &= p^2q(-1+p+\tfrac{1}{2}q+\tfrac{1}{2}q^2-\tfrac{1}{2}pq)\\
\delta_2 &= pq^2(-1+\tfrac{1}{2}p+q+\tfrac{1}{2}p^2-\tfrac{1}{2}pq)\\
\delta_3 &= pqr^2\\
\epsilon_1 &= p^2[\tfrac{1}{2}p(1-p)^2+5/4\ q^2r]\\
\epsilon_2 &= 1/4\ p^2q^2r+\tfrac{1}{2}p^2q^3\\
\epsilon_3 &= \tfrac{1}{2}p^2r^2(1-p+2q)\\
\Phi_1 &= 1/4\ p^2q^2r+\tfrac{1}{2}p^3q^2\\
\Phi_2 &= q^2[\tfrac{1}{2}q(1-q)^2+5/4\ p^2r]\\
\Phi_3 &= \tfrac{1}{2}q^2r^2(1+2p-q),
\end{aligned}$$

with $r = 1 - p - q$.

Corrective Functions

An alternative method of producing a C' approximating function over a triangular network is to start with the C^0 approximating function given by (10) and to add corrective terms which will increase the continuity of the function to C'. The main properties of these corrective functions are that they must vanish on the perimeter of the triangle and that they must reduce the normal derivative of the function along the sides of the triangle from quadratic to linear form. At the same time, of course, they must not destroy the continuity of the function and slope inside the triangular element.

One set of corrective functions which is in common use (Zienkiewicz, (14) p. 117) consists of

$$A(p,q,r) = \frac{pq^2r^2}{(1-q)(1-r)},\quad B(p,q,r) = \frac{p^2qr^2}{(1-p)(1-r)},$$
$$C(p,q,r) = \frac{p^2q^2r}{(1-p)(1-q)} \tag{14b}$$

In fact, Dupuis and Goel, have shown that C' continuity is obtained if the right hand sides of (10a), (10b), and (10c) are supplemented by

$$2[-A+(2-3\frac{L_3}{L_2}\cos\theta_1)B + (2-3\frac{L_2}{L_3}\cos\theta_1)C] \tag{15a}$$

$$\tfrac{1}{2}[(\xi_3-\xi_2)A + \left\{3\xi_2+5\xi_3+6\eta_2\frac{L_3}{L_2}\sin\theta_1\right\}B + \left\{-5\xi_2-3\xi_3+6\eta_3\frac{L_2}{L_3}\sin\theta_1\right\}C] \tag{15b}$$

and

$$\tfrac{1}{2}[\eta_3-\eta_2)A+\left\{3\eta_2+5\eta_3-6\xi_2\frac{L_3}{L_2}\sin\theta_1\right\}B-\left\{5\eta_2+3\eta_3+6\xi_3\frac{L_2}{L_3}\sin\theta_1\right\}C] \tag{15c}$$

respectively, where $\theta_i (i = 1,2,3)$ is the angle subtended at the vertex, and $L_i (i = 1,2,3)$ is the length of the side opposite the vertex P_i. The supplements for the other functions in formula (10) are obtained by cyclic interchanges. It should be remembered that the functions A, B and C given by (14) take part in the cyclic interchanges.

Another set of corrective functions obtained by Clough and Tocher (5) requires each triangle to be divided into three small triangles having the centroid of the triangle as a common vertex. The corrective functions are

$$
\begin{aligned}
\alpha(p,q,r) &= \begin{cases} p^2(3-5p)-6pqr & \text{in } \Pi_2 G \Pi_3 \\ q^2(q-3r) & \text{in } \Pi_3 G \Pi_1 \\ r^2(r-3q) & \text{in } \Pi_1 G \Pi_2 \end{cases} \\
\beta(p,q,r) &= \begin{cases} p^2(p-3r) & \text{in } \Pi_2 G \Pi_3 \\ q^2(3-5q)-6pqr & \text{in } \Pi_3 G \Pi_1 \\ r^2(r-3p) & \text{in } \Pi_1 G \Pi_2 \end{cases} \\
\gamma(p,q,r) &= \begin{cases} p^2(p-3q) & \text{in } \Pi_2 G \Pi_3 \\ q^2(q-3p) & \text{in } \Pi_3 G \Pi_1 \\ r^2(3-5r)-6pqr & \text{in } \Pi_1 G \Pi_2 \end{cases}
\end{aligned}
\tag{16}
$$

This time C' continuity is obtained by adding (15a), (15b), and (15c) to the right hand sides of (10a), (10b), and (10c) respectively with A, B and C replaced by $-1/6\ \alpha$, $1/6\ \beta$ and $-1/6\ \gamma$ respectively.

2. THE RECTANGLE

Rectangular type regions, i.e. regions with sides parallel to the x- and y-axes, occur in many problems in physics and engineering. Consequently, the rectangular element is of considerable importance, and basis functions will now be constructed for it.

Cubic Hermites

In this section we now consider the bicubic polynomial written in the form

$$g(x,y) = \sum_{r=0}^{3} \sum_{s=0}^{3} \alpha_{r,s} x^r y^s \tag{17}$$

over the unit square $0 \le x,y \le 1$. The coefficients $\alpha_{r,s}$ $(0 \le r,s \le 3)$ can be found uniquely in terms of the values of $g, \frac{\partial g}{\partial x}, \frac{\partial g}{\partial y}, \frac{\partial^2 g}{\partial x \partial y}$ at the four corners of the square. Substituting the values of the coefficients into (17) and rearranging the right hand side leads to the result

$$\begin{aligned} g(x,y) = {} & (1-x)^2(1+2x)(1-y)^2(1+2y)g_1+(1-x)^2x(1-y)^2(1+2y)(\tfrac{\partial g}{\partial x})_1 \\ & +(1-x)^2(1+2x)(1-y)^2y(\tfrac{\partial g}{\partial y})_1+(1-x)^2x(1-y)^2y(\tfrac{\partial^2 g}{\partial x \partial y})_1 \\ & +x^2(3-2x)(1-y)^2(1+2y)g_2+x^2(x-1)(1-y)^2(1+2y)(\tfrac{\partial g}{\partial x})_2 \\ & +x^2(3-2x)(1-y)^2y(\tfrac{\partial g}{\partial y})_2+x^2(x-1)(1-y)^2y(\tfrac{\partial^2 g}{\partial x \partial y})_2 \\ & +(1-x)^2(1+2x)y^2(3-2y)g_3+(1-x)^2xy^2(3-2y)(\tfrac{\partial g}{\partial x})_3 \\ & +(1-x)^2(1+2x)y^2(y-1)(\tfrac{\partial g}{\partial y})_3+(1-x)^2xy^2(y-1)(\tfrac{\partial^2 g}{\partial x \partial y})_3 \\ & +x^2(3-2x)y^2(3-2y)g_4+(x-1)^2x^2y^2(3-2y)(\tfrac{\partial g}{\partial x})_4 \\ & +x^2(3-2x)y^2(y-1)(\tfrac{\partial g}{\partial y})_4+x^2(x-1)y^2(y-1)(\tfrac{\partial^2 g}{\partial x \partial y})_4 \end{aligned} \tag{18}$$

It is an easy matter to see how the result (18) can be modified to give the required Hermite bicubic interpolating function over any rectangular element of the original rectangular type region. The approximating function over the complete region is then obtained. This time there are four basis functions corresponding to each node of the rectangular array. For an internal node, i.e. a node not on the boundary of the rectangular type region, each basis function has a support of four rectangular elements. For a node on the boundary, but not at a corner, the support is two rectangular elements, and for the corner nodes one rectangular element. Over the complete rectangular region the piecewise bicubic approximating function gives $C^{1,1}$ continuity. *

* $C^{i,j}$ continuity for a function $u(x,y)$ means all derivatives $\frac{\partial^{\ell+m}u}{\partial^{\ell}x\partial^{m}y}$ $(0 \leq \ell \leq i,\ 0 \leq m \leq j)$ continues over the complete region.

Cubic Splines

The cubic spline function in one dimension with compact support of 4h, first suggested by Schoenberg (12) is often written as

$$M(x) = \frac{1}{6}\,\delta^4\{(\frac{x}{h} - i)_+^3\} \tag{19}$$

where δ is the usual central difference operator, and the constant $\frac{1}{6}$ is chosen so that

$$\int_{-\infty}^{+\infty} M(x)dx = 1$$

The constant $\frac{1}{4}$ in (15) was chosen to make

$$B_i(i) = 1.$$

In rectangular type regions subdivided into rectangular elements, we consider the Schoenberg splines M(x), given by (19), and M(y), given by

$$M(y) = \frac{1}{6}\,\delta^4\{(\frac{y}{h} - j)_+^3\}.$$

The tensor product of M(x) and M(y) gives rise to a bell type function with a support of sixteen rectangular elements. This is the basis function for cubic splines at the node $x = ih$, $y = jh$ provided the latter is neither on the boundary of the region, nor adjacent to the boundary. For such nodes, special basis functions have to be constructed, unless of course the problem being solved has natural boundary conditions. The overall approximating function in this case has $C^{2,2}$ continuity.

It is perhaps worth mentioning that in the unlikely event of $C^{4,4}$ continuity being required of the approximating function on a rectangular region, the Schoenberg quintic splines

$$M(x) = \frac{1}{5!}\,\delta^6\,(\frac{x}{h} - i)_+^5$$

can be used in a manner similar to that of the cubic splines of the last section. The tensor products have a support of thirty-six rectangular elements, which makes the biquintic splines rather difficult to handle.

In concluding this short section on the rectangular element it is worth pointing out that whereas the support of the spline ($4\ell^2$ elements) increases with the order of the spline $(2\ell-1)$, where $\ell = 1,2,3,\ldots$, the support of the Hermite function remains constant at four elements, irrespective of the order of the Hermite function $(2\ell-1)$. The spline of course has greater continuity $C^{2(\ell-1),2(\ell-1)}$ as against $C^{\ell-1,\ell-1}$ for the Hermite function.

3. THE QUADRILATERAL

It might be thought that quadrilaterals are better mesh units than triangles because the overall grid is simplified. For example, a triangular network can always be simplified by combining the triangles in twos or fours to form quadrilaterals. Unfortunately, however, it is impossible to find a polynomial in x and y which reduces to linear form along the four sides of an arbitrary quadrilateral, and so it is not obvious how one can construct a piecewise function in x and y which has C^0 continuity over a quadrilateral network.

To find such a function, we introduce the following notation. Let $P_1P_2P_3P_4$ be a quadrilateral in the x,y-plane, with P_5P_6 as the exterior diagonal. If the lines P_1P_7 and P_2P_8 are perpendiculars of unit length to the x,y - plane, then

$$P_i \equiv (x_i, y_i, 0)$$

for $i = 1,2,\ldots,6$ and

$$P_i \equiv (x_i, y_i, 1)$$

for $i = 7,8$. We denote by C_{ijk} the constants

$$\begin{vmatrix} 1 & x_i & y_i \\ 1 & x_j & y_j \\ 1 & x_k & y_k \end{vmatrix}$$

and by D_{ij} the linear forms

$$\begin{vmatrix} 1 & x & y \\ 1 & x_i & y_i \\ 1 & x_j & y_j \end{vmatrix} .$$

Finally, we denote by Π_1, Π_2, Π_3 and Π_4 respectively, the linear forms

$$- zC_{124} + D_{24} - D_{14},$$
$$- zC_{123} + D_{23} - D_{13},$$
$$- zC_{234} + D_{34}$$

and

$$- zC_{134} + D_{34},$$

where z is the coordinate perpendicular to the x,y - plane. The planes $P_4P_7P_8$, $P_3P_7P_8$, $P_3P_4P_8$ and $P_3P_4P_7$ are then given by $\Pi_1 = 0$, $\Pi_2 = 0$, $\Pi_3 = 0$ and $\Pi_4 = 0$ respectively.

The surface

$$\alpha\Pi_1\Pi_3 - \beta\Pi_2\Pi_4 = 0 \quad (\alpha,\beta \text{ arbitrary constants})$$

passes through the points P_3,P_4,P_7 and P_8. The equation can be written in full as

$$(\alpha C_{234}C_{124}-\beta C_{134}C_{123})z^2 + [\{C_{234}(D_{14}-D_{24})-C_{124}D_{34}\}\alpha - \{C_{134}(D_{13}-D_{23})-C_{123}D_{34}\}\beta]z - D_{34}[(D_{14}-D_{24})\alpha - (D_{13}-D_{23})\beta] = 0. \quad (20)$$

Since

$$D_{14} - D_{24} = D_{12} - C_{124}$$

and

$$D_{13} - D_{23} = D_{12} - C_{123},$$

equation (20) becomes

$$(\alpha C_{734}C_{124} - \beta C_{134}C_{123})z^2 + [(\alpha C_{234} - \beta C_{134})D_{12} -$$
$$(\alpha C_{124} - \beta C_{123})D_{34} - (\alpha C_{234}C_{124} - \beta C_{134}C_{123})]z - \tag{21}$$
$$D_{34}[(\alpha-\beta)D_{12} - (\alpha C_{124} - \beta C_{123})] = 0$$

It is easily seen that for any values of the parameters α and β, if $z = 0$ then $D_{34} = 0$ and if $z = 1$ then $D_{12} = 0$. The pair of equations $z = 0$ and $D_{34} = 0$, gives the line P_3P_4 and the pair of equations $z = 1$ and $D_{12} = 0$, gives the line P_7P_8; hence the surface given by (21), contains these two lines for any values of the parameters α and β.

Particular examples of surfaces given by (21) are

(i) $\alpha C_{234}C_{124} = \beta C_{134}C_{123}$. Here the quadratic term dissappears and we are left with the surface

$$z = \frac{[\frac{1}{2}\{(C_{234}+C_{134})-(C_{124}+C_{123})\}D_{12}+C_{124}C_{123}]D_{34}}{C_{234}C_{134}D_{12}+C_{124}C_{123}D_{34}} \tag{22}$$

This gives z as a rational function of x and y, with the product of two linear forms in the numerator and a single linear form in the denominator.

(ii) $\alpha = \beta$. This time (21) reduces to

$$\tfrac{1}{2}[(C_{234}+C_{134})-(C_{124}+C_{123})]z^2 - [D_{34}+D_{12}+\tfrac{1}{2}\{(C_{234}+C_{134})$$
$$-(C_{124}+C_{123})\}]z + D_{34} = 0 \tag{24}$$

It is easily seen that if z = constant, (24) becomes a linear equation in x and y, and so each plane z = constant intersects the surface in a straight line.

Isoparametric Coordinates

It is a simple step from the ideas of the last few paragraphs to isoparametric coordinates for the quadrilateral. (Irons (7), Zienkiewicz (15)). Suppose we now consider surfaces containing the line P_2P_3 and the line through P_7 parallel to P_1P_4. Cyclic interchanges in equation (21) lead to surfaces of the form

$$(\alpha'C_{134}C_{123}-\beta'C_{234}C_{124})z^2 - [(\alpha'C_{123}-\beta'C_{234})D_{14} + (\alpha'C_{134}-\beta'C_{124})D_{23} + (\alpha'C_{134}C_{123}-\beta'C_{234}C_{124})]z + D_{23}[(\alpha'-\beta')D_{14} + (\alpha'C_{134}-\beta'C_{124})] = 0 \quad (25)$$

where α',β' are arbitrary constants. Equations (21) and (25) can be used to define local isoparametric coordinates p and q, for the quadrilateral $P_1P_2P_3P_4$ such that

$$(\alpha'C_{134}C_{123}-\beta'C_{234}C_{124})p^2 - [(\alpha'C_{123}-\beta'C_{234})D_{14} + (\alpha'C_{134}-\beta'C_{124})D_{23} + (\alpha'C_{134}C_{123}-\beta'C_{234}C_{124})]p + D_{23}[(\alpha'-\beta')D_{14} + (\alpha'C_{134}-\beta'C_{124})] = 0 \quad (26a)$$

and

$$(\alpha C_{234}C_{124}-\beta C_{134}C_{123})q^2 + [(\alpha C_{234}-\beta C_{134})D_{12} - (\alpha C_{124}-\beta C_{123})D_{34} - (\alpha C_{234}C_{124}-\beta C_{134}C_{123})]q - D_{34}[(\alpha-\beta)D_{12} - (\alpha C_{124}-\beta C_{123})] = 0 \quad (26b)$$

Different choices of the parameter pairs α',β' and α,β produce different sets of isoparametric coordinates for the quandrilateral. Each set has $p = 0,1$ along the sides P_2P_3, P_1P_4 and $q = 0,1$ along the sides P_3P_4, P_2P_1 respectively.

As an example of the isoparametric coordinates p and q given by (26), consider the quadrilateral $P_1P_2P_3P_4$, where $p = 0,1$ along the sides P_2P_3, P_1P_4 and $q = 0,1$ along the sides P_3P_4, P_2P_1. The formulae connecting the global coordinates x,y with the local isoparametric coordinates p,q are

$$x = pqx_1+(1-p)qx_2+(1-p)(1-q)x_3+p(1-q)x_4 \tag{27a}$$

$$y = pqy_1+(1-p)qy_2+(1-p)(1-q)y_3+p(1-q)y_4 \tag{27b}$$

These formulae can be manipulated to give

$$\begin{aligned}&[(x_1-x_2)(y_4-y_3)-(y_1-y_2)(x_4-x_3)]p^2 + [(y_1-y_2-y_4+y_3)\\ &(x-x_3)-(x_1-x_2-x_4+x_3)(y_1-y_3)+(y_4-y_3)(x_2-x_3)-(x_4-x_3)\\ &(y_2-y_3)]p + [(y_1-y_3)(x-x_3)-(x_2-x_3)(y-y_3)] = 0\end{aligned} \tag{28a}$$

and

$$\begin{aligned}&[(x_1-x_4)(y_2-y_3)-(y_1-y_2)(x_2-x_3)]q^2 + [(y_1-y_2-y_4+y_3)\\ &(x-x_3)-(x_1-x_2-x_4+x_3)(y-y_3)+(y_2-y_3)(x_4-x_3)-(x_2-x_3)\\ &(y_4-y_3)]q + [(y_4-y_3)(x-x_3)-(x_4-x_3)(y-y_3)] = 0\end{aligned} \tag{28b}$$

We consider now the case of a quadrilateral with four side points. (i.e. one point on each side). In the (p,q) plane, the quadrilateral becomes the unit square. The transformation formulae are

$$\begin{aligned} x = {} & pq(2p-1)(2q-1)x_1+(1-p)q(2q-1)(1-2p)x_2 \\ & +(1-p)(1-q)(1-2p)(1-2q)x_3+p(1-q)(1-2q)(2p-1)x_4 \\ & +4pq(1-p)(2q-1)x_5+4q(1-p)(1-q)(1-2p)x_6 \\ & +4(1-p)(1-q)p(1-2q)x_7+4pq(2p-1)(1-q)x_8 \\ & +16pq(1-p)(1-q)x_9, \end{aligned} \tag{29}$$

y similarly.

The point P_9 is included so that the full biquadratic polynomial in p and q can be used. This is similar to the case of the triangle where the centroid is included so that the full cubic polynomial can be utilised. The sides of the quadrilateral can be made straight by suitable positioning of the side nodes P_5, P_6, P_7 and P_8. In particular if these are mid points of the respective sides and P_9 is taken to be the centroid of the quadrilateral, then formulae (29) reduce to formulae (27).

The internal node P, can be removed in a variety of ways. For example, from (29), the term in p^2q^2 can be eliminated by choosing

$$x_9 = \tfrac{1}{2}(x_5+x_6+x_7+x_8) - \tfrac{1}{4}(x_1+x_2+x_3+x_4)$$

$$y_9 = \tfrac{1}{2}(y_5+y_6+y_7+y_8) - \tfrac{1}{4}(y_1+y_2+y_3+y_4),$$

and formulae (29) become

$$\begin{aligned} x = {} & pq(2p+2q-3)x_1+(1-p)q(-2p+2q-1)x_2 \\ & +(1-p)(1-q)(1-2p-2q)x_3+p(1-q)(2p-2q-1)x_4 \\ & +4pq(1-p)x_5+4q(1-p)(1-q)x_6 \\ & +4pq(1-p)(1-q)x_7+4pq(1-q)x_8, \end{aligned} \tag{30}$$

y similarly.

This eight node quadrilateral was analysed by Jordan (8) as an element for use in solving problems involving plane stress or strain.

Although no mention has been made in this section of the form of the interpolating function on the quadrilateral, it is convenient to choose it to be the same as that given by the geometry definition. Indeed this is the essential characteristic of <u>isoparametric elements</u>. Consequently for the quadrilateral, from (27) the interpolating function is given by $U(x,y) = pqU_1 + (1-p)qU_2 + (1-p)(1-q)U_3 + p(1-q)U_4$, and for the quadrilateral with side points, it is given by (29), or (30), with U replacing x.

Finally for the quadrilateral with eight side points (i.e. two points on each side), the full bicubic polynomial in p and q can be used for interpolation provided four internal nodes are included. The internal nodes are removed by eliminating the terms in p^2q^2, p^3q^2, p^2q^3, and p^3q^3, leading to the interpolating function

$$\begin{aligned} U(x,y) = {} & \tfrac{9}{2}(p^2+q^2-p-q+\tfrac{2}{9})[pqU_1+(1-p)qU_2+(1-p)(1-q)U_3 \\ & +p(1-q)U_4]+\tfrac{9}{2}pq(1-p)[(3p-1)U_5+(2-3p)U_6] \\ & +\tfrac{9}{2}(1-p)q(1-q)[(3q-1)U_7+(2-3q)U_8]+\tfrac{9}{2}p(1-p) \\ & (1-q)[(2-3p)U_9+(3p-1)U_{10}]+\tfrac{9}{2}pq(1-q)[(2-3q)U_{11} \\ & + (3q-1)U_{12}] \end{aligned} \qquad (31)$$

4. THE TETRAHEDRON

Lagrangian Interpolation

The relevant interpolating polynomial of degree m can be expressed as

$$U(x,y,z) = \sum_{r=1}^{\frac{1}{6}(m+1)(m+2)(m+3)} U_r p_r^{(m)}(x,y,z)$$

where the summation is over all nodes of the tetrahedron, and $p_r^{(m)}(x,y,z)$ is a polynomial basis function taking the value unity at the node in question and zero at all other nodes. The first three cases of this family of basis functions are

(i) m = 1. Here there are four basis functions $p_r^{(1)}(r = 1,2,3,4)$ which for convenience we call p,q,r, and s respectively. These basis functions are related to the cartesian coordinates x,y, and z through the formula

$$\begin{bmatrix} 1 \\ x \\ y \\ z \end{bmatrix} = \begin{bmatrix} 1 & 1 & 1 & 1 \\ x_1 & x_2 & x_3 & x_4 \\ y_1 & y_2 & y_3 & y_4 \\ z_1 & z_2 & z_3 & z_4 \end{bmatrix} \begin{bmatrix} p \\ q \\ r \\ s \end{bmatrix} . \tag{32}$$

The geometric significance of this relationship is that

p = 0 on the plane $P_2P_3P_4$, p = 1 at the vertex P_1,
q = 0 " " " $P_3P_4P_1$, q = 1 " " " P_2
r = 0 " " " $P_4P_1P_2$, r = 1 " " " P_3
s = 0 " " " $P_1P_2P_3$, s = 1 " " " P_4,

where of course

$$p + q + r + s = 1.$$

The tetrahedron is referred to as the <u>standard</u> tetrahedron.

It is useful to realise that

$$\begin{vmatrix} 1 & 1 & 1 & 1 \\ x_1 & x_2 & x_3 & x_4 \\ y_1 & y_2 & y_3 & y_4 \\ z_1 & z_2 & z_3 & z_4 \end{vmatrix} \qquad 6V,$$

where V is the volume of the tetrahedron, and so the matrix in (31) can be inverted to give the coordinates p,q,r, and s in terms of x,y, and z.

(ii) $m = 2$. There are ten nodes, four corner and six mid-side. The basis functions are

Corner nodes

$p(2p-1)$ at 1 2,3,4 similarly

Mid-side nodes

$4pq$ at mid point of 12 others similarly

(iii) $m = 3$. There are twenty nodes, four corner, twelve at the points of trisection of the six sides, and four at the centroids of the four faces. The basis functions at these nodes are

Corner nodes

$\frac{1}{2}p(3p-1)(3p-2)$ at 1 2,3,4 similarly

Points of trisection of sides

$\frac{9}{2}pq(3p-1)$ at first point of trisection of 12 others similarly

Mid face nodes

27 pqr at centroid of face 123. others similarly

Hermitian Interpolation

An interesting variant of the full third order polynomial in the three variables x,y, and z is obtained by interpolating U(x,y) and all its first derivatives at the four vertices together with U(x,y) at the centroids of each of the four faces. This is an example of Hermitian interpolation in three dimensions and is extremely complicated. It is unlikely that this element will be used extensively, but nevertheless, we shall quote the interpolating polynomial in the form (see Ciarlet and Wagschal (2)).

$$\begin{aligned} \Pi_3(x,y,z) = &\sum_{i=1}^{4}[3p_i^2 - 2p_i^3 - 7\sum_{1\le i<j<k\le 4} p_i p_j p_k]U_i \\ &+ 27\sum_{1\le i<j<k\le 4} p_i p_j p_k U_{ijk} \\ &+ \sum_{\substack{i,j=1\\ i\ne j}}^{4} p_i^2 p_j \left\{D_{\underline{ij}} U_i\right\} \\ &- \sum_{i,j,k}^{1} p_i p_j p_k \left\{D_{\underline{ij}} U_i\right\} \end{aligned} \tag{33}$$

where we write p_1, p_2, p_3, p_4, for p,q,r,s, U_{ijk} is the value of U at the centroid of the triangular face ijk; $D_{\underline{ij}} U_i$ means the first derivative of U at the vertex i with respect to the side ij multiplied by the length of the side ij;

and $\sum_{i,j,k}^{1}$ means sum over all $1 \le i,j,k \le 4$ for which $i \neq j$, $j \neq k$ and $k \neq i$.

5. THE HEXAHEDRON

The element having six faces is known as the hexahedron. In general it is a better element in three dimensions than the tetrahedron. The minimum munber of vertices required for a hexahedron is eight. If coordinates p,q, and r are introduced where each coordinate takes the values 0,1 on a pair of opposite faces, the hexahedron becomes a unit cube in (p,q,r) space. These coordinates are known as the natural coordinates for the hexahedral element.

It is a simple matter to see that the transformation formulae from the natural coordinates p,q,r to the global cartesian coordinates x,y,x are given by

$$\begin{aligned} x = {} & pqrx_1+(1-p)qrx_2+(1-p)(1-q)rx_3+p(1-q)rx_4+p(1-r)x_5 \\ & +(1-p)q(1-r)x_6+(1-p)(1-q)(1-r)x_7+p(1-q)(1-r)x_8 \end{aligned} \qquad (34)$$

$\left.\begin{matrix} y \\ z \end{matrix}\right\}$ similarly.

In the case of isoparametric elements, the interpolating polynomial is given by (34) with U replacing x. It is trilinear in form and can be written as

$$\begin{aligned} U(x,y,z) = {} & U_7+(U_8-U_7)p+(U_6-U_7)q+(U_3-U_7)r \\ & +(U_5-U_6+U_7-U_8)pq+(U_2-U_3-U_6+U_7)qr \\ & +(U_4-U_3+U_7-U_8)rp+(U_1-U_2+U_3-U_4-U_5+U_6-U_7+U_8)pqr. \end{aligned}$$

Higher order tetrahedra are obtained in a similar manner. For example, if a triquadratic interpolation form in p,q, and r is used, we require twenty-seven nodes in the tetrahedron. These are made up of eight corner nodes, twelve side nodes, six face nodes and a node inside the tetrahedron. Hence the interpolating polynomial for the case of the nodes symmetrically placed in the cube is

$$U(x,y,z) = \sum_{i=1}^{8} \alpha_i U_i + \sum_{i=q}^{20} \beta_i U_i + \sum_{i=21}^{26} \gamma_i U_i + \delta_{27} U_{27} \qquad (35)$$

where

$\alpha_1 = pqr(2p-1)(2q-1)(2r-1)$ $\qquad \alpha_2,\ldots,\alpha_8$ similarly

$\beta_9 = 4pqr(1-p)(2q-1)(2r-1)$ $\qquad \beta_{10},\ldots,\beta_{20}$ "

$\gamma_{21} = 16pqr(1-p)(1-q)(2r-1)$ $\qquad \gamma_{22},\ldots,\gamma_{26}$ "

$\delta_{27} = 64pqr(1-p)(1-q)(1-r).$

The nodes numbered 9, 21, and 27 are located in the tetrahedron at $(p,q,r) = (\frac{1}{2},1,1)$, $(\frac{1}{2},\frac{1}{2},1)$, and $(\frac{1}{2},\frac{1}{2},\frac{1}{2})$ respectively. The six centre face nodes and the node at the centre of the tetrahedron can be removed by eliminating the terms in p^2q^2, p^2r^2, r^2p^2, p^2q^2r, p^2qr^2, pq^2r^2, and $p^2q^2r^2$ from (35). The resulting interpolating formula for the twenty node tetrahedron is

$$U(x,y,z) = \sum_{i=1}^{8} \alpha_i' U_i + \sum_{i=9}^{20} \beta_i' U_i \qquad (36)$$

where

$\alpha_1' = pqr(2p+2q+2r-5)$ $\qquad \alpha_2',\ldots,\alpha_8'$ similarly

$\beta_9' = 4pqr(1-p)$ $\qquad \beta_{10}',\ldots,\beta_{20}'$ "

Tricubic interpolation in a hexahedron requires sixty-four nodes. These are located with respect to the cube as follows; eight at the corner points, twenty-four at the points of trisection of the twelve sides, twenty-four on the fix faces (four symmetrically placed in each face) and eight symmetrically placed inside the cube. This is a rather complicated hexahedral element and a more popular version is obtained by considering the first thirty-two nodes mentioned viz. the eight corner nodes together with the twenty-four nodes at the points of trisection of the sides. The resulting interpolating formula for this thirty-two node tetrahedron is

$$U(x,y,z) = \sum_{i=1}^{8} \alpha_i U_i + \sum_{i=9}^{32} \beta_i U_i \tag{37}$$

where

$$\alpha_1 = \frac{9}{2}pqr[(p^2+q^2+r^2)-(p+q+r)+\frac{2}{9}], \quad \alpha_2,\ldots,\alpha_8 \text{ similarly}$$

$$\beta_9 = \frac{9}{2}pqr(1-p)(3p-1). \qquad \beta_{10},\ldots,\beta_{32} \quad ''$$

The node numbered 1 is located at $(p,q,r) = (1,1,1)$ and the node numbered 9 at $(p,q,r) = (\frac{2}{3},1,1)$.

A general comparison of the relative merits of tetrahedral and hexahedral elements for the analysis of elastic solids can be found in a survey article by Clough (4).

6. CURVED BOUNDARIES

So far basis functions have been constructed in the main for networks with straight sides. In real problems in two and three dimensions, however, boundaries and interfaces are often curved. It is the purpose of this section to derive basis functions for networks composed of elements with curved sides (two dimensions) or curved surfaces (three dimensions). The curved element was introduced into structural analysis by Ergatondis, Irons, and Zienkiewicz (6) and reference to it can be found in Zienkiewicz (15).

Two Dimensions

Elements with straight sides, usually triangles or quadrilaterals, are perfectly satisfactory if the domain has a polygonal boundary. If some part of the boundary is curved, however, elements with at least one curved side are desirable. This is also the case when curved material interfaces are present in the region.

Initially we consider the triangular element with two straight and one curved side. This element together with triangles with straight sides can deal adequately with most plane problems involving curved boundaries and interfaces.

Triangle With One Curved Side

The triangle $P_1P_2P_3$ is considered in the (x,y) plane with $\ell(x,y) = 0$ and $m(x,y) = 0$ the equations of the straight sides P_2P_3 and P_3P_1 respectively. The equation of the curved side passing through P_1 and P_2 is $f(x,y) = 0$, and ℓ,m are normalised so that

$$\ell(x_1,y_1) = m(x_2,y_2) = 1,$$

where (x_i,y_i) are the coordinates of the point $P_i(i = 1,2,3)$. In the (ℓ,m) plane, the triangle becomes $P_1'P_2'P_3'$ where $P_1' \equiv (1,0)$, $P_2' \equiv (0,1)$ and $P_3' \equiv (0,0)$ and the curve $f(\ell,m) = 0$ is normalised so that

$$f(0,0) = 1$$

The connecting linear relations are given by

$$\ell = \frac{(y_2-y_3)x-(x_2-x_3)y-(y_2x_3-x_2y_3)}{2\Delta}$$

$$m = \frac{-(y_1-y_3)x+(x_1-x_3)y+(y_1x_3-x_1y_3)}{2\Delta}$$

where

$$2\Delta = \begin{vmatrix} 1 & x_1 & y_1 \\ 1 & x_2 & y_2 \\ 1 & x_3 & y_3 \end{vmatrix}$$

Consider now the family of surfaces $z(\ell,m) = 0$, which intersects the (ℓ,m) plane in the curve $f(\ell,m) = 0$, and is given by the equation

$$z(\alpha z+\beta\ell+\gamma m+\delta) + f(\ell,m) = 0, \tag{38}$$

where α, β, γ and δ are parameters. If we impose the conditions

(i) $z = 1$ at $\ell = m = 0$

(ii) $z = 1 - \ell$ at $m = 0$

(iii) $z = 1 - m$ at $\ell = 0$,

then (38) becomes

$$\alpha z^2 + [\alpha(\ell+m-1) + 1 - \frac{f(\ell,o)}{1-\ell} - \frac{f(o,m)}{1-m}]z + f(\ell,m) = 0 \tag{39}$$

where z is now playing the role of an <u>isoparametric</u> coordinate, introduced in the previous section. In particular if $\alpha = 0$, (39) reduces to

$$z = \frac{f(\ell,m)}{\frac{f(o,m)}{1-m} + \frac{f(\ell,o)}{1-\ell} - 1} \tag{40}$$

If, in addition, the curve $f(\ell,m) = 0$ is a conic, then

$$f(\ell,m) \equiv a\ell^2 + b\ell m + cm^2 - (a+1)\ell - (c+1)m + 1 = 0,$$

and (40) reduces to

$$z = \frac{a\ell^2 + b\ell m + cm^2 - (a+1)\ell - (c+1)m+1}{1-a\ell-cm} \tag{41}$$

Now the approximating function over a triangular element is often expressed in the Lagrangian form

$$U_\Delta(x,y) = \sum_{i\in S} U_i W_i(x,y);W_i(x_j,y_j)=\begin{cases}1(j=i)\\0(j\neq i)\end{cases} j\in S \tag{42}$$

where S is a selected set of points on the triangle, and $W_i(x,y)$ is a wedge (part basis) function associated with the node i. If it is required for the wedges to form a basis for linear approximation over each element, then the conditions

$$\sum W_i = 1, \sum x_i W_i = x, \sum y_i W_i = y \tag{43a}$$

are imposed. In terms of the coordinates (ℓ,m) these give

$$\sum W_i = 1, \sum \ell_i W_i = \ell, \sum m_i W_i = m. \tag{43b}$$

Since the wedge function W_3 must satisfy (39) and the complete set (43), the minimum number of wedge functions for a triangular element with two straight and one curved side is four. These functions are given by

$$\begin{aligned}
W_1 &= \frac{(1-M)\ell+Lm-L}{1-L-M} + \frac{L}{1-L-M} W_3 \\
W_2 &= \frac{M\ell+(1-L)m-M}{1-L-M} + \frac{M}{1-L-M} W_3 \\
W_4 &= \frac{1-\ell-m}{1-L-M} - \frac{1}{1-L-M} W_3,
\end{aligned} \tag{44}$$

with W_3 given by (39), and (L,M) any node P_4' on the curved side $P_1'P_2'$. The wedge functions W_1, W_2, W_3 and W_4 correspond to P_1', P_2', P_3' and P_4' respectively.

Connection With Isoparametric Coordinates

At this stage it is interesting to show the connection between the surfaces z = 0 given by (39) and isoparametric coordinates for a triangle with one curved and two straight sides. First consider a general curved triangle. The isoparametric coordinates p,q, and r are self explanatory and it is easy to see that p + q + r = 1. The node Π_i in the (p,q) plane corresponds to the node P_i in the (x,y) plane, where i = 1,2,...,6. The unique quadratic interpolating polynomial in the (p,q) plane is

$$U(p,q) = p(2p-1)U_1+q(2q-1)U_2+r(2r-1)U_3+4pqU_4+4qrU_5+4rpU_6 \tag{45a}$$

with isoparametric elements, the interpolation formula includes a linear polynomial in x and y, and so

$$x = p(2p-1)x_1+q(2q-1)x_2+r(2r-1)x_3+4pqx_4+4qrx_5+4rpx_6 \tag{45b}$$

y similarly.

For the special case of a triangle with two straight and one curved side, put $x_5 = \frac{1}{2}(x_2+x_3)$, $x_6 = \frac{1}{2}(x_3+x_1)$, and y similarly. Formulae (45) simplify to give

$$\begin{aligned} \ell &= 2(2R-1)pq + p \\ m &= 2(2R-1)pq + q \end{aligned} \tag{46}$$

where P_4' is given by $\ell = m = R$. The isoparametric coordinates, p,q, and r (≡1-p-q) can be isolated in turn from (46) to give

$$p^2 + [m-\ell+ \frac{1}{2(2R-1)}]p - \frac{\ell}{2(2R-1)} = 0, \tag{47a}$$

$$q^2 + [\ell-m+ \frac{1}{2(2R-1)}]q - \frac{m}{2(2R-1)} = 0, \tag{47b}$$

and

$$r^2 - \frac{4R-1}{2R-1} r + [\frac{2R-\ell-m}{2R-1} - (\ell-m)^2] = 0 \tag{47c}$$

respectively. If we put r = 0 in (46) the curved side is given by

$$f(\ell,m) \equiv 1 - \frac{\ell+m}{2R} - \frac{2R-1}{2R} (\ell-m)^2 = 0. \tag{48}$$

It is easily seen that (39) is equivalent to (47c) when $f(\ell,m)$ is given by (48) and $\alpha = \frac{2R-1}{R}$. The curve given by (48) is of course parabola.

Forbidden Elements

One of the difficulties in dealing with curved elements using isoparametric coordinates arises from the vanishing of the Jacobian of the transformation from the (ℓ,m) coordinate system to the (p,q) system. For example, if we consider the formulae

$$\begin{aligned} \ell &= 2(2L-1)pq + p \\ m &= 2(2M-1)pq + q \end{aligned} \tag{49}$$

which are equivalent to (46) with P_4' given by $\ell = L$, $m \doteq M$, the Jacobian of the transformation is

$$J \equiv \begin{vmatrix} \frac{\partial \ell}{\partial p} & \frac{\partial \ell}{\partial q} \\ \frac{\partial m}{\partial p} & \frac{\partial m}{\partial q} \end{vmatrix} = 1 + 2(2M-1)p + 2(2L-1)q$$

The Jacobian is positive for $0 \le p \le 1 - q$, $0 \le q \le 1$ provided (L,M) lies in the open region of the (ℓ,m) plane bounded by the lines $\ell = m = 1/4$. For other positions of the point (L,M) in the positive quadrant of the (ℓ,m) plane, including the lines $\ell = 1/4$, $m = 1/4$, the Jacobian either vanishes or is negative for certain values of (p,q), (Jordan (8)), and so isoparametric coordinates cannot be used to deal with curved elements in these "forbidden" cases. This is because results calculated in terms of the isoparametric coordinates (p,q) cannot be transferred back to the (ℓ,m) plane because of the vanishing of the Jacobian of the transformation somewhere in the element.

Instead of using isoparametric coordinates, we can deal directly in terms of ℓ and m by using (39). In the example discussed previously where the curved side is given by (48) and $\alpha = \frac{2R-1}{2R}$, (39) reduces to (47). This equation will have real roots provided

$$F(\ell,m;R) \equiv \left(\ell+m+\frac{1}{2(2R-1)}\right)^2 - 4\ell m \ge 0. \tag{50}$$

It is easy to see that for a fixed value of R the function $F(\ell,m;R)$ has no maximum or minimum inside the element, or indeed anywhere in the (ℓ,m) plane. Consequently the smallest value of $F(\ell,m;R)$ will occur on the boundary of the element for all values of R. In fact, using (48) it follows that $F(\ell,m;R) = \frac{4R-1^2}{4R-2} > 0$ for all R on the curved side, and of course F is always positive from (50) on $\ell = 0$ and $m = 0$. The condition (50) is thus satisfied for all values of R and for all points (ℓ,m) in the element.

When isoparametric coordinates are used in this example, values of R in the range $0 < R < 1/4$ are forbidden. This does not seem to be the case when (39) is used. However by simple geometrical considerations, it can be shown that for $0 < R < 1/4$, the curved boundary intersects the ℓ and m axes at points between the origin and the unit points.

The Quadratic Case

It has already been shown that, when the triangle with two straight and one curved sides has a Lagrange interpolating function which is a special case of (45a), the isoparametric coordinates p,q,-and r are given by (47). Each of these coordinates represents a special case of the quantity z which satisfies (39), and so p,q, and r are linear along the straight sides of the triangle in the (ℓ,m) plane. Hence it follows that the interpolating function (45a) is quadratic along the straight sides in the (ℓ,m) plane and so there is no difficulty in obtaining C^0 continuity in a network composed of straight sided triangles, each with a quadratic polynomial interpolating function, adjoined by triangles with two straight and one curved sides round the perimeter of the region.

In the quadratic case it is also possible to obtain the interpolating function directly in the (ℓ,m) plane without using isoparametric coordinates. This is done by introducing extra nodal points P_5' and P_6' at the mid points of the sides $\ell = 0$ and $m = 0$ respectively. If W_1 is any solution of (39), then the part basis function at node P_1' in the quadratic case can be taken to be

$$(W_3)_Q = (1-2\ell-2m)W_3$$

The required wedge function at node P_5' is zero along the straight side 361 and the curved side 142, is quadratic along the straight side 253 and takes the value unity at P_5'. Such a function is

$$(W_5)_Q = \frac{4mf(\ell,m)}{1-a\ell-cm} \quad ,$$

in the case where

$$f(\ell,m) \equiv a\ell^2+b\ell m+cm^2-(1+a)\ell-(1+c)m+1 \tag{51}$$

with

$$b = -\frac{1}{LM}(aL^2+cM^2-(1+a)L-(1+c)M+1).$$

Similarly

$$(W_6)_Q = \frac{4\ell f(\ell,m)}{1-a\ell-cm}.$$

The remaining wedge functions in the quadratic case with $f(\ell,m)$ given by (51) are obtained from (43b), and are given by

$$(W_1)_Q = \ell(1-\frac{2f}{1-a\ell-cm}) - \frac{L}{1-L-M}[(1-\ell-m) - \frac{2f(\ell+m)}{1-a\ell-cm} -$$

$$(1-2\ell-2m)W_3]$$

$$(W_2)_Q = m(1 - \frac{2f}{1-a\ell-cm}) - \frac{M}{1-L-M}[(1-\ell-m) - \frac{2f(\ell+m)}{1-a\ell-cm} -$$

$$(1-2\ell-2m)W_3]$$

$$(W_4)_Q = \frac{1}{1-L-M}[(1-\ell-m) - \frac{2f(\ell+m)}{1-a\ell-cm} - (1-2\ell-2m)W_3]$$

In the special case $\alpha = a = c = 0$, these formulae reduce to

$$(W_1)_Q = \ell(1 - \frac{m}{M} - 2f)$$

$$(W_2)_Q = m(1 - \frac{\ell}{L} - 2f)$$

$$(W_3)_Q = (1 - 2\ell - 2m)f$$

$$(W_4)_Q = \ell m/LM$$

$$(W_5)_Q = 4mf$$

$$(W_6)_Q = 4\ell f$$

where $f \equiv \frac{L+M-1}{LM}\ell m - \ell - m + 1.$

The Cubic Case

The unique cubic interpolating polynomial in the (p,q) is

$$U(p,q) = \tfrac{1}{2}p(3p-1)(3p-2)U_1 + \tfrac{1}{2}q(3q-1)(3q-2)U_2 + \tfrac{1}{2}r(3r-1)(3r-2)U_3 + \frac{9}{2}pq(3p-1)U_4 + \frac{9}{2}pq(3q-1)U_5 + \frac{9}{2}qr(3q-1)U_6 + \frac{9}{2}qr(3r-1)U_7 + \frac{9}{2}rp(3r-1)U_8 + \frac{9}{2}rp(3p-1)U_9 + 27pqr\,U_{10} \quad (53)$$

with $r = 1 - p - q$, where U_i $(i = 1,2,\ldots,10)$ is the value of U at the node Π_i $(i = 1,2,\ldots,10)$. With isoparametric elements, (53) holds with U replaced by x and y (or ℓ and m) respectively, and in the special case of a triangle with P_2P_3 and P_3P_1 straight sides and with P_6, P_7 and P_8, P_9 points of trisection of P_2P_3 and P_3P_1 respectively, these formulae reduce to

$$\ell = p[1+\frac{9}{2}(q^2-q)]\ell_1 + q[1+\frac{9}{2}(p^2-p)]\ell_2 + (p+q-1)(9pq-1)\ell_3 + \frac{9}{2}pq[(3p-1)\ell_4 + (3q-1)\ell_5] + 27pq(1-p-q)\ell_{10}.$$

m = similarly.

Since $\ell_1 = 1$, $\ell_2 = \ell_3 = 0$, and $m_2 = 1$, $m_1 = m_3 = 0$, we get

$$\ell = p + \frac{9}{2}pq(6\ell_{10}-\ell_4-\ell_5-1) + \frac{27}{2}p^2q(\ell_4-2\ell_{10}) + \frac{27}{2}pq^2(\ell_5-2\ell_{10}+1/3)$$

$$m = q + \frac{9}{2}pq(6m_{10}-m_4-m_5-1) + \frac{27}{2}p^2q(m_4-2m_{10}+1/3) + \frac{27}{2}pq^2(m_5-2m_{10}).$$

It follows that p and q are linear along the straight sides of the triangle in the (ℓ,m) plane, and so the interpolating function (53) is cubic along the straight sides in the (ℓ,m) plane, leading to C^o continuity in a network composed of straight sided triangles, each with a cubic polynomial interpolating function, adjoined by triangles with two straight and one curved side round the perimeter of the region.

Again in the cubic case, it is possible to obtain the interpolating function directly in the (ℓ,m) plane without using isoparametric coordinates. In order to obtain wedge functions which are cubic along the straight sides, we introduce extra nodal points P_5' and P_6' on $\ell = 0$ and P_7' and P_8' on $m = 0$. These nodes are at the points of trisection of the sides. A simple extension of the ideas used in the quadratic case leads to the wedge functions

$$(W_3)_c = (1 - 3\ell - 3m)(1 - 3/2\ \ell - 3/2\ m)\ W_3$$

$$(W_5)_c = \frac{9/2\ m(3m-1)f}{1 - a\ell - cm}$$

$$(W_6)_c = \frac{-\ 9m(3/2\ m-1)f}{1 - a\ell - cm}$$

$$(W_7)_c = \frac{-\ 9\ell(3/2\ \ell-1)f}{1 - a\ell - cm}$$

$$(W_8)_c = \frac{9/2\ \ell(3\ell-1)f}{1 - a\ell - cm}\ ,$$

where

$$f \equiv a\ell^2 + b\ell m + cm^2 - (1+a)\ell - (1+c)m + 1.$$

The remaining wedge functions in the cubic case are given by the equations

$$(W_1)_c+(W_2)_c+(W_4)_c = 1 - (W_3)_c-(W_5)_c-(W_6)_c-(W_7)_c-(W_8)_c$$

$$(W_1)_c \qquad +L(W_4)_c = \ell - 1/3\ (W_7)_c - 2/3\ (W_8)_c$$

$$(W_2)_c+M(W_4)_c = m - 2/3\ (W_5)_c - 1/3\ (W_6)_c,$$

which solve to give

$$(W_1)_c = \ell\ (1 - \frac{8/3 - 7/2\ \ell}{1 - a\ell - cm}\ f) - L(W_4)_c$$

$$(W_2)_c = m\ (1 - \frac{8/3 - 7/2\ m}{1 - a\ell - cm}\ f) - M(W_4)_c$$

$$(W_4)_c = \frac{1}{1-L-M}\ [(1-\ell-m) - 9/2\ \frac{\ell(1-\ell)+m(1-m)}{1 - a\ell - cm}\ f$$

$$- (1-3\ell-3m)(1-3/2\ \ell - 3/2 \cdot m)\ W_3].$$

In the particular case, $\alpha = a = c = 0$, these formulae reduce to

$$(W_3)_c = (1-3\ell-3m)(1 - 3/2\ \ell - 3/2\ m)\ f$$

$$(W_1)_c = \ell\ [1 - \frac{m}{M} + (\frac{9L}{1-L-M}\ m + 7/2\ \ell - 8/3)\ f]$$

$$(W_2)_c = m\ [1 - \frac{\ell}{L} + (\frac{9L}{1-L-M}\ \ell + 7/2\ m - 8/3)\ f]$$

$$(W_4)_c = \frac{\ell m}{LM} - \frac{9\ell m}{1-L-M}\ f$$

$$(W_5)_c = 9/2\ m\ (3m - 1)\ f$$

$$(W_6)_c = - 9m(3/2\ m - 1)\ f$$

$$(W_7)_c = - 9\ell(3/2\ \ell - 1)\ f$$

$$(W_8)_c = 9/2\, \ell\, (3\ell-1)\, f$$

where

$$f \equiv \frac{L+M-1}{LM}\, \ell m - \ell - m + 1$$

Quadrilateral With One Curved Side

In the case of a quadrilateral with one intermediate node on each side, it follows from (30) that for an isoparametric element, the interpolating function is given by

$$\begin{aligned} U = {} & pq(2p+2q-3)U_1 + (1-p)q\,(-2p+2q-1)U_2 + (1-p)(1-q)(1-2p-2q)U_3 \\ & + p(1-q)(2p-2q-1)U_4 + 4pq(1-p)U_5 + 4q(1-p)(1-q)U_6 \\ & + 4p(1-p)(1-q)U_7 + 4pq(1-q)U_8. \end{aligned}$$

For the case of one curved and three straight sides where P_6, P_7 and P_8 are the mid points of the sides P_2P_3, P_3P_4 and P_4P_1 respectively, the point transformation formulae (30) reduce to

$$\begin{aligned} \ell &= p + [-2-A+4L]pq + [2+2A-4L]\, p^2q \\ m &= q + [-3-B+4M]pq + [2+2B-4M]\, p^2q \end{aligned} \qquad (55)$$

where

$$\ell = \frac{(y_2-y_3)x - (x_2-x_3)y + (x_2y_3-x_3y_2)}{\begin{vmatrix} 1 & x_2 & y_2 \\ 1 & x_3 & y_3 \\ 1 & x_4 & y_4 \end{vmatrix}}$$

and

$$m = \frac{(y_3-y_4)x - (x_3-x_4)y + (x_3y_4-x_4y_3)}{\begin{vmatrix} 1 & x_2 & y_2 \\ 1 & x_3 & y_3 \\ 1 & x_4 & y_4 \end{vmatrix}}$$

From (55), the isoparametric coordinate p isolates to give

$$Tp^3 + [Z + (T\ell-Ym)]p^2 + [1 + Xm - Z\ell]\, p - \ell = 0,$$

and the curve q = 1 is given by

$$f(\ell,m) \equiv [T\ell+Y(1-m)]^2 + (T+XT-YZ)[Z\ell+(1+X)(1-m)] = 0, \quad (56)$$

where

$$X = -2 - A + 4L, \qquad Y = 2 + 2A - 4L,$$
$$Z = -3 - B + 4M, \qquad T = 2 + 2B - 4M.$$

This curve is of course a parabola. Hence if isoparametric coordinates are used, as defined by the point transformation formulae (30), the curved side is replaced by the parabolic arc, whose equation is given by (56).

For the case of a quadrilateral with two intermediate nodes on each side, it follows from (55) that the point transformation formulae reduce to

$$\ell = p + 9/2\,[(2/9\,A-L+2N)pq - (A-4L+5N+1)p^2q + (A-3L+3N+1)p^3q]$$
$$m = q + 9/2\,[2/9\,B-M+2Q-11/9)pq - (B-4M+5Q-2)p^2q + (B-3M+3Q-1)p^3q] \quad (57)$$

The isoparametric coordinate q can be eliminated to give a quartic in p, and the curve q = 1 is a quartic in ℓ and m.

REFERENCES

1. Birkhoff, G. Proc. Nat. Acad. Sci. 68 p. 1162 (1971).

2. Ciarlet, P.G. and Wagschal, C. Numer. Math. 17 p. 84 (1971)

3. Ciarlet, P.G. and Raviart, P.A. Arch. Rat. Mech. Anal. 46, p. 177 (1972).

4. Clough, R.W. Address to the American Society of Civil Engineers (1971).

5. Clough, R.W. and Tocher, J.L. Proc. 1st Conf. Matrix Methods in Structural Mechanics Wright-Patterson AFB Ohio (1965).

6. Ergatondis, I., Irons, B.M., and Zienkiewicz, O.C. Int. J. Solids Structures 4, p. 31 (1968).

7. Irons, B.M. Conf. on Use of digital computers in Structural Engineering Newcastle (1966).

8. Jordan, W.B., A.E.C. Research and Development Report KAPL - M-7112 (1970).

9. McLeod, R.Y.J. and Mitchell, A.R. J. Inst. Maths. Applics. 10, p. 382 (1972).

10. Mitchell, A.R. Conf. on the Finite Element Method Brunel (1972).

11. Mitchell, A.R., Phillips, G. and Wachspress, E.L. J. Inst. Maths. Applics. 8, p. 260 (1971).

12. Schoenberg, I.J. Approximations with special emphasis on spline functions Academic Press (1969).

13. Wachspress, E.L. J. Inst. Maths. Applics. 11, p. 83 (1973).

14. Zienkiewicz, O.C. The Finite Element Method in Structural and Continuum Mechanics McGraw Hill (1967).

15. Zienkiewicz, O.C. The Finite Element Method in Egineering Science McGraw Hill (1971).

16. Zlamal, M. Numer. Math. 14 p. 394 (1970).

Quadratic Interpolating Splines: Theory and Applications

W. J. Kammerer, G. W. Reddien, and R. S. Varga*

Georgia Inst. of Technology | Vanderbilt University | Kent State University

§1. Introduction.

Our basic aim in this paper is to study the projectional properties of quadratic interpolatory splines. In §3, we show that in Theorem 3.1 that the linear interpolating projection operator P_{Δ} from $C^{-1}[a, b]$, the space of all bounded functions on $[a, b]$, to $Sp(2, \Delta)$, the space of quadratic splines on $[a, b]$, is bounded, for any partition Δ of $[a, b]$. In §4, we develop global interpolation error bounds for quadratic spline interpolation. Finally, in §5, we develop local interpolation error bounds.

§2. Notation.

For $-\infty < a < b < +\infty$ and for any positive integer $N \geq 2$, let

*Research supported in part by the U.S. Atomic Energy Commission under Grant AT(11-1)-2075.

(2.1) Δ: $a = x_0 < x_1 < x_2 < \cdots < x_N = b$

denote a partition of [a, b] with knots x_i. The collection of all such partitions of [a, b] is called $\mathcal{P}(a, b)$. We define $\overline{\Delta} = \max\{(x_{i+1} - x_i): 0 \le i \le N-1\}$ and $\underline{\Delta} = \min\{(x_{i+1} - x_i): 0 \le i \le N-1\}$ for each partition of the form (2.1). For any real number σ with $\sigma \ge 1$, $\mathcal{P}_\sigma(a, b)$ then denotes the subset of $\mathcal{P}(a, b)$ of all partitions Δ for which $\overline{\Delta} \le \sigma\underline{\Delta}$. In particular, $\mathcal{P}_1(a, b)$ is the collection of all <u>uniform</u> partitions of [a, b].

If π_m denotes the collection of all real algebraic polynomials of degree at most m, then for any nonnegative integer n, the <u>polynomial spline space</u> $Sp(n, \Delta)$ is defined as usual by

(2.2) $$Sp(n, \Delta) = \{s(x): s \in C^{n-1}[a, b], s(x) \in \pi_n \text{ for } x \in (x_i, x_{i+1}), i = 0, 1, \cdots, N-1\},$$

where, for the case n = 0, $C^{-1}[a, b]$ denotes the set of bounded functions on [a, b]. We remark that $Sp(n, \Delta)$ is a finite-dimensional linear subspace of $C^{n-1}[a, b]$. In particular, we shall be concerned here with the <u>quadratic spline space</u> $Sp(2, \Delta) \subset C^1[a, b]$, which can be seen to be of dimension N + 2. For additional notation, we set $h_i = x_i - x_{i-1}$, $i = 1, 2, \cdots, N$, and, for g(x) defined on [a, b], we set $g_i = g(x_i)$, $i = 0, 1, \cdots, N$, and $g_{i+1/2} = g(x_i + \frac{h_{i+1}}{2})$, $i = 0, 1, \cdots, N-1$.

Given any $s \in Sp(2, \Delta)$, the restriction of s on any subinterval $[x_{i-1}, x_i]$ determined by Δ is simply the unique quadratic polynomial interpolating the values s_{i-1}, $s_{i-1/2}$, and s_i, respectively in the points x_{i-1}, $x_{i-1/2}$, and x_i. Comparing the restrictions of s on $[x_{i-1}, x_i]$ and on $[x_i, x_{i+1}]$, and using the fact (cf. (2.2)) that s is continuously differentiable at x_i, it is easily verified (cf. Marsden [11]) that

$$(2.3)\quad a_i s_{i-1} + 3s_i + c_i s_{i+1} = 4a_i s_{i-1/2} + 4c_i s_{i+1/2}, 1 \le i \le N-1,$$

where

$$(2.4)\quad a_i \equiv \frac{h_{i+1}}{h_i + h_{i+1}};\ c_i \equiv \frac{h_i}{h_i + h_{i+1}} = 1 - a_i,\ 1 \le i \le N-1,$$

so that $0 < a_i, c_i < 1$ and $a_i + c_i = 1$, $1 \le i \le N-1$. In a completely similar way, one can show that

$$(2.5)\quad c_i\ Ds_{i-1} + 3Ds_i + a_i\ Ds_{i+1} = 8(x_{i+1/2} - s_{i-1/2})/(h_i + h_{i+1}),$$

$$1 \le i \le N-1,$$

where $Ds_j \equiv \frac{ds(x_i)}{dx}$. The identities of (2.3) and (2.5) will be used in developing error bounds for quadratic spline interpolation.

§3. Sp(2, Δ) Interpolation.

Given any $f \in C^{-1}[a, b]$, let $s \in Sp(2, \Delta)$ be its (unique) interpolant defined by the N + 2 conditions

$$(3.1)\quad s_0 = f_0,\ s_{i+1/2} = f_{i+1/2},\ 0 \leq i \leq N-1,\ s_N = f_N.$$

From (2.3) and (3.1), this implies that the values $\{s_i\}_{i=1}^{N-1}$ satisfy the N-1 linear equations

$$(3.2)\quad a_i s_{i-1} + 3s_i + c_i s_{i+1} = 4a_i f_{i-1/2} + 4c_i f_{i+1/2},\ 1 \leq i \leq N-1,$$

with $s_0 = f_0$ and $s_N = f_N$. With (2.4), the associated $(N - 1) \times (N - 1)$ coefficient matrix for the s_i's from (3.2) is evidently strictly diagonally dominant and hence non-singular (cf. [15, p. 23]). This proves that the values $\{s_i\}_{i=1}^{N-1}$ are uniquely determined. But, as s is the unique quadratic interpolation on $[x_{i-1}, x_i]$ of the values s_{i-1}, $s_{i-1/2}$, and s_i in the points x_{i-1}, $x_{i-1/2}$, and x_i, then s is also uniquely determined.

Our interest now is in the projectional properties of the interpolants of all $f \in C^{-1}[a, b]$, i.e., for any bounded function defined on [a, b]. The following result is stated in Marsden [11], and, because its proof is both short and elementary, we include it below.

Theorem 3.1. Given any $f \in C^{-1}[a, b]$ and given any $\Delta \in \mathcal{P}(a,b)$, let s be the unique interpolant of f in Sp(2, Δ), in the sense of (3.1). Then,

$$(3.3) \qquad \|s\|_{L_\infty[a,b]} \le 2\,\|f\|_{L_\infty[a,b]}.$$

Proof. We first show (by a typical diagonal dominance argument) that

$$(3.4) \qquad \max\{|s_i| : 0 \le i \le N\} \le 2\,\|f\|_{L_\infty[a,b]}.$$

From (2.3), $3s_i = 4a_i f_{i-1/2} + 4c_i f_{i+1/2} - a_i s_{i-1} - c_i s_{i+1}$, $1 \le i \le N-1$. Taking absolute values and using the fact (cf. (2.4)) that $a_i + c_i = 1$, then

$$(3.5)\ 3|s_i| \le 4\|f\|_{L_\infty[a,b]} + \max\{|s_j| : 0 \le j \le N\},\ 1 \le i \le N-1.$$

Let $|s_j| = \max\{|s_i| : 0 \le i \le N\}$. If $j = 0$ or $j = N$, then from (3.1), $\max\{|s_j| : 0 \le j \le N\} = |s_j| = |f_j| \le \|f\|_{L_\infty[a,b]}$, which certainly implies (3.4). If, on the other hand, j satisfies $1 \le j \le N-1$, then choosing $i = j$ in (3.5) gives $2\max\{|s_j| : 0 \le j \le N\} \le 4\,\|f\|_{L_\infty[a,b]}$, the desired result of (3.4).

With (3.4), we now establish (3.3). Consider any subinterval $[x_{i-1}, x_i]$ determined by Δ. By a change of scale, we may assume, without loss of generality, that $x_{i-1} = 0$, $x_i = 1$, $h_i = 1$. On this subinterval, the interpolant s is a quadratic polynomial, explicitly given by

$$s(x) = [s_{i-1}(1 - x) - s_i \cdot x](1 - 2x) + 4s_{i-1/2}x(1 - x), x \in [0,1].$$

Thus, using (3.4) and (3.1), a short calculation shows that

$$|s(x)| \le 2\|f\|_{L_\infty[a,b]} \cdot |1 - 2x| + 4\|f\|_{L_\infty[a,b]} \cdot x(1 - x) \le 2\|f\|_{L_\infty[a,b]}$$

for all $x \in [0, 1]$. As this holds for an arbitrary subinterval of Δ, then (3.3) is valid. Q.E.D.

If $\mathcal{P}_\Delta: C^{-1}[a, b] \to Sp[2, \Delta]$ denotes the linear projection mapping of any $f \in C^{-1}[a, b]$ into its unique $Sp(2, \Delta)$-interpolant $s \equiv P_\Delta f$ in the sense of (3.1), the result of Theorem 3.1 directly gives us the following

<u>Corollary 3.2</u>. For any partition $\Delta \in \mathcal{P}(a, b)$, the projection mapping P_Δ satisfies

$$(3.6) \quad \|P_\Delta\|_\infty \equiv \sup\{\|P_\Delta f\|_{L_\infty[a,b]} : \|f\|_{L_\infty[a,b]} \le 1\} \le 2.$$

Following Marsden [11] in the periodic case, it can also be shown that the upper bound of (3.6) is <u>sharp</u>, i.e., for any ϵ with $0 < \epsilon < 2$, there is a partition Δ of $[a, b]$ for which $\|P_\Delta\|_\infty > 2 - \epsilon$.

The interesting feature of the above corollary is that the quadratic spline interpolation mappings P_Δ are uniformly bounded, <u>independent</u> of any assumption on the partitions Δ, unlike the case for, say, cubic spline interpolation mappings (cf. [1], [3], [4], [9], [10] and [12]).

§4. <u>Global Error Bounds for Quadratic Spline Interpolation</u>.

We consider now global error bounds for quadratic spline

interpolation. As in §3, let P_Δ denote the linear projection mapping of $C^{-1}[a, b]$ onto $Sp(2, \Delta)$, and as usual, let

$\omega(g, \delta, I) \equiv \sup\{|g(x) - g(y)| : |x - y| \le \delta \text{ and } x,y \in I\}$

denote the L_∞-modulus of continuity of g with respect to the interval I. If $I = [a, b]$, we write simply $\omega(g, \delta)$ for $\omega(g, \delta, [a, b])$.

Because P_Δ is a (bounded linear) projector on $Sp(2, \Delta)$, i.e., $w = P_\Delta w$ for any $w \in Sp(2, \Delta)$, it is well known that, for any $f \in C^{-1}[a, b]$ and for any $w \in Sp(2, \Delta)$,

$$\|f - P_\Delta f\|_{L_\infty[a,b]} = \|f - w - P_\Delta f + P_\Delta w\|_{L_\infty[a,b]} =$$

$$\|(I - P_\Delta)(f - w)\|_{L_\infty[a,b]} \le (1 + \|P_\Delta\|_\infty)\|f - w\|_{L_\infty[a,b]}.$$

Thus, as this inequality is valid for any $w \in Sp(2, \Delta)$, it follows from (3.6) that

$$(4.1) \quad \|f - P_\Delta f\|_{L_\infty[a,b]} \le 3 \inf\{\|f - w\|_{L_\infty[a,b]} : w \in Sp(2, \Delta)\}.$$

We now use a special case of results from de Boor and Fix [2]. Recalling the definition of $Sp(n, \Delta)$ in (2.2), it follows from [2] that if $f \in C^k(I)$ and $I \subseteq [a, b]$ and with $0 \le k \le n$, then

$$(4.2) \quad \inf\{\|D^j(f - w)\|_{L_\infty[I]} : w \in Sp(n, \Delta)\} \le c_{j,k}(\overline{\Delta})^{k-j}\omega(D^k f, \overline{\Delta}, I),$$

$$0 \le j \le k,$$

where the constants $c_{j,k}$ are independent of f, and independent of the partition Δ if $0 \le j \le \left[\frac{n+1}{2}\right]$, while for $\left[\frac{n+1}{2}\right] < j \le n$, the constants $c_{j,k}$ depend only on the local mesh ratio M of (4.3) below, determined from a somewhat larger interval covering I. Specifically, if knots x_{τ_1} and x_{τ_2} of the partition Δ are chosen so that $[x_{\tau_1}, x_{\tau_2}] \supseteq I$, then

$$(4.3)\ M \equiv \max\left\{\left(\frac{x_{i+1} - x_i}{x_{j+1} - x_j}\right) : |i - j| = 1 \text{ and } \tau_1 - n \le i,j \le \tau_2 + n - 1\right\}.$$

Thus, coupling the inequality in (4.1) with the inequality in (4.2) for the case $n = 2$, $j = 0$, and $I = [a, b]$, directly yields

<u>Theorem 4.1</u>. Given any $f \in C^k[a, b]$ with $0 \le k \le 2$, and given any $\Delta \in \mathcal{P}(a, b)$, then

$$(4.4)\quad \|f - P_\Delta f\|_{L_\infty[a,b]} \le C_{0,k}(\bar{\Delta})^k\, \omega(D^k f, \bar{\Delta}),$$

where the constants $C_{0,k}$ are independent of f and Δ.

To derive global derivative error bounds for $f - P_\Delta f$, the above technique using (4.2) can again be used. Specifically, suppose we wish to bound $\|D(f - P_\Delta f)\|_{L_\infty[a,b]}$ for $f \in C^k[a, b]$, $1 \le k \le 2$. Writing $f(x) = f(a) + \int_a^x g(t)\,dt$, so that $g(x) = Df(x)$, define $P_{\Delta,1}g$ as

$$(P_{\Delta,1}g)(x) \equiv D(P_\Delta f)(x).$$

It is easily verified that $P_{\Delta,1}$ is a linear projector in $C[a, b]$ with range $Sp(1, \Delta)$, and one readily establishes as in Corollary 3.2, via a diagonal dominance argument based on (2.5), that $\|P_{\Delta,1}\|_\infty \le 2$ for any $\Delta \in \mathcal{P}(a, b)$. Thus, as in (4.1),

$$\|D(f - P_\Delta f)\|_{L_\infty[a,b]} = \|g - P_{\Delta,1}g\|_{L_\infty[a,b]} \le 3 \cdot \inf\left\{\|g - w\|_{L_\infty[a,b]} : w \in Sp(1, \Delta)\right\}.$$

Applying the inequality of (4.2) for the case $j = 0$, $n = 1$, and $I = [a, b]$ to the right-hand side of the above inequality directly gives

Theorem 4.2. Given any $f \in C^k[a, b]$ and $1 \le k \le 2$, and given any $\Delta \in \mathcal{P}(a, b)$, then

$$(4.5)\qquad \|D(f - P_\Delta f)\|_{L_\infty[a,b]} \le C_{1,k}(\bar{\Delta})^{k-1}\,\omega(D^k f, \bar{\Delta}),$$

where the constants $C_{1,k}$ are independent of f and Δ.

To obtain error bounds for $\|D^2(f - P_\Delta)\|_{L_\infty[a,b]}$, we make use of the following well-known Markov inequality for polynomials (cf. Lorentz [7, p. 40]): for any $p \in \pi_m$, and for any finite real numbers α and β with $\beta > \alpha$,

$$(4.6)\qquad \|Dp\|_{L_\infty[\alpha,\beta]} \le \frac{2m^2}{(\beta - \alpha)}\|p\|_{L_\infty[\alpha,\beta]}.$$

For any $f \in C^2[a, b]$ and for any $\Delta \in \mathcal{P}_\sigma(a, b)$, let $\hat{w} \in Sp(2, \Delta)$ satisfy (4.2), i.e.,

$$(4.7) \quad \|D^j(f - \hat{w})\|_{L_\infty[a,b]} \leq c_{j,2}(\overline{\Delta})^{2-j}\ \omega(D^2 f, \overline{\Delta}), \quad 0 \leq j \leq 2.$$

On any subinterval (x_i, x_{i+1}) of Δ, $\hat{w}$ and $P_\Delta f$ are evidently in π_2, and applying (4.6) twice gives

$$(4.8) \quad \|D^2(\hat{w} - P_\Delta f)\|_{L_\infty[x_i,x_{i+1}]} \leq \frac{16}{(\underline{\Delta})^2} \|\hat{w} - P_\Delta f\|_{L_\infty[x_i,x_{i+1}]},$$

since $(x_{i+1} - x_i) \geq \underline{\Delta}$. But, as

$$(4.9) \quad \|\hat{w} - P_\Delta f\|_{L_\infty[x_i,x_{i+1}]} \leq \|\hat{w} - P_\Delta f\|_{L_\infty[a,b]} \leq$$
$$\|\hat{w} - f\|_{L_\infty[a,b]} + \|f - P_\Delta f\|_{L_\infty[a,b]},$$

and as each term on the right is bounded above from (4.4) and (4.7) by $C_{0,2}(\overline{\Delta})^2\ \omega(D^2 f, \overline{\Delta})$, it follows for $\Delta \in \mathcal{P}_\sigma(a, b)$ that

$$(4.10) \quad \|D^2(\hat{w} - P_\Delta f)\|_{L_\infty[a,b]} \leq C_{0,2}\ \sigma^2\ \omega(D^2 f, \overline{\Delta}).$$

Hence, since $\|D^2(f - P_\Delta f)\|_{L_\infty[a,b]} \leq \|D^2(f - \hat{w})\|_{L_\infty[a,b]} + \|D^2(\hat{w} - P_\Delta f)\|_{L_\infty[a,b]}$, then applying (4.7) and (4.10) gives

<u>Theorem 4.3</u>. Given any $f \in C^2[a, b]$ and given any $\Delta \in \mathcal{P}_\sigma(a, b)$, then

$$(4.11) \quad \|D^2(f - P_\Delta f)\|_{L_\infty[a,b]} \leq C_{2,2}\ \omega(D^2 f, \overline{\Delta}),$$

where $C_{2,2}$ is independent of f, but is dependent on σ.

We remark that Marsden [11] has, by different direct means, given explicit upper bounds for the constants $C_{j,k}$ appearing in (4.4), (4.7), and in (4.11). In the case of (4.11), the results of [11] are derived under the stronger hypothesis that $\Delta \in \mathcal{P}_1(a, b)$, i.e., Δ is a uniform partition of [a, b]. Note, however, that the corresponding result $j = k = 2$ of (4.2) of de Boor and Fix [2] is derived under the weaker hypothesis of locally bounded mesh ratios (cf. (4.3)).

§5. Local Rates of Convergence.

We now develop local rates of convergence for quadratic spline interpolants. The argument parallels techniques found in Swartz and Varga [14] and Kammerer and Reddien [5], in that a given bounded function f is approximated by a smooth function g, and then, using global interpolation error bounds for g developed in §4, along with the exponential damping properties of the off-diagonal entries of the inverse of the coefficient matrix associated with (2.3), local interpolation error are then obtained for f.

For notation to be used throughout this section, we shall consider any bounded function f defined on [a, b], i.e., $f \in C^{-1}[a, b]$, with $f \in C^k[a', b']$, where $0 \le k \le 2$ and where $[a', b'] \equiv I' \underset{\neq}{\subset} [a, b]$. We shall then focus our attention on a fixed closed interval $[\alpha, \beta] \subset [a', b']$, where, for a given δ with $0 < \delta < 1/2$,

$$(5.1)\quad \begin{cases} \alpha = a' + \delta(b' - a') \text{ whenever } a' \neq a;\ \alpha = a \text{ otherwise, and} \\ \beta = b' - \delta(b' - a') \text{ whenever } b' \neq b;\ \beta = b \text{ otherwise.} \end{cases}$$

We then consider partitions $\Delta \in \mathcal{P}(a, b)$ for which

$$(5.2)\quad 2\overline{\Delta} \leq \delta.$$

With f a given function in $C^{-1}[a, b]$ with $f \in C^k[a', b']$ where $0 \leq k \leq 2$, let g denote any fixed function such that

$$(5.3)\quad \begin{cases} \text{i)} & g \in C^k[a, b], \\ \text{ii)} & g \equiv f \text{ on } [a', b'] = I', \\ \text{iii)} & \omega(D^k g, t) \equiv \omega(D^k f, t, I') \text{ for any } 0 < t \leq b' - a'. \end{cases}$$

For example, with $g \equiv f$ on $[a', b']$ from (5.3ii), such a function can be obtained simply by defining g on $[a, a']$ to be the unique element in π_k with $D^j g(a') = D^j f(a')$, $0 \leq j \leq k$, with an analogous definition for g in $[b', b]$.

If $[t]$ denotes as usual the greatest integer less than or equal to t, we state

<u>Theorem 5.1</u>. Given any $f \in C^{-1}[a, b]$ with $f \in C^k[a', b']$ where $0 \leq k \leq 2$, let $P_\Delta f$ denote the quadratic spline interpolant of f in $Sp(2, \Delta)$, in the sense of (3.1). Then, for any $\Delta \in \mathcal{P}(a, b)$ with $2\overline{\Delta} \leq \delta$,

$$(5.4)\quad \|f - P_\Delta f\|_{L_\infty[\alpha,\beta]} \leq C_{0,k}(\overline{\Delta})^k\, \omega(D^k f, \overline{\Delta}, I') + \frac{45\|g-f\|_{L_\infty[a,b]}}{8\cdot 3^{[\delta/\overline{\Delta}]}},$$

where g satisfies (5.3), and where the constants $C_{0,k}$ are those of Theorem 4.1.

As an immediate consequence of Theorem 5.1, we have

Corollary 5.2. With the hypotheses of Theorem 5.1, for every $\epsilon > 0$, there exists an $\eta > 0$ such that for any $\Delta \in \mathcal{P}(a, b)$ with $\bar{\Delta} < \eta$,

$$\|f - P_\Delta f\|_{L_\infty[\alpha,\beta]} \le (C_{0,k} + \epsilon)(\bar{\Delta})^k \, \omega(D^k f, \bar{\Delta}, I').$$

In particular, if $D^k f$ is constant on I', then

$$\|f - P_\Delta f\|_{L_\infty[\alpha,\beta]} \le \frac{45\|g - f\|_{L_\infty[a,b]}}{8 \cdot 3^{[\delta/\bar{\Delta}]}}.$$

We next state

Theorem 5.3. Given any $f \in C^{-1}[a, b]$ with $f \in C^k[a', b']$ where $1 \le k \le 2$, then for any $\Delta \in \mathcal{P}_\sigma(a, b)$ with $2\bar{\Delta} \le \delta$,

$$(5.5)\quad \|D^j(f - P_\Delta f)\|_{L_\infty[\alpha,\beta]} \le C_{j,k}(\bar{\Delta})^{k-j}\omega(D^k f, \bar{\Delta}, I') + \gamma_{j,k}\frac{\|g-f\|_{L_\infty[a,b]}}{3^{[\delta/\bar{\Delta}]}}$$

for $1 \le j \le k$, where the constants $C_{j,k}$ are those of Theorem 4.2 and 4.3, and the constants $\gamma_{j,k}$ are independent of f, but dependent on σ.

With the hypotheses of Theorems 5.1 and 5.3 and the fixed function g satisfying (5.3), the triangle inequality gives us for $0 \le j \le k$ that

(5.6) $$\|D^j(f-P_\Delta f)\|_{L_\infty[\alpha,\beta]} \le \|D^j(f-g)\|_{L_\infty[\alpha,\beta]} + \|D^j(g-P_\Delta g)\|_{L_\infty[\alpha,\beta]} + \|D^j(P_\Delta(g-f))\|_{L_\infty[\alpha,\beta]}.$$

From the definition of g in (5.3ii), the first term on the right side of (5.6) is necessarily zero, while the second term can be bounded above from the results of §4 by

$$\|D^j(g - P_\Delta g)\|_{L_\infty[\alpha,\beta]} \le \|D^j(g - P_\Delta g)\|_{L_\infty[a,b]}$$

$$\le C_{j,k}(\bar{\Delta})^{k-j}\ \omega(D^k g,\ \bar{\Delta}) = C_{j,k}(\bar{\Delta})^{k-j}\ \omega(D^k f,\ I',\ \bar{\Delta}),$$

the last equality following from (5.3iii). Thus, it remains to bound above the last term on the right of (5.6). If $s \equiv P_\Delta(g - f)$, then s satisfies the hypotheses of the following lemma with $M = \|g - f\|_{L_\infty[a,b]}$. The results of (5.8)-(5.10) of Lemma 5.4 then establish Theorems 5.1 and 5.3.

Lemma 5.4. Given a positive constant M, and given $\Delta \in \mathcal{P}(a, b)$ satisfying (5.2), let $s \in Sp(2, \Delta)$ satisfy:

(5.7)
i) $s_{i-1/2} = 0$ whenever $x_{i-1/2} \in [a', b']$; $s_i = 0$ whenever $x_i \in [a', b']$ and $i = 0$ or $i = N$,
ii) $|s_{i-1/2}| \le M$ for all $1 \le i \le N$, and $|s_0| \le M$, $|s_N| \le M$.

Then,

(5.8) $$\|s\|_{L_\infty[\alpha,\beta]} \le \frac{45}{8} M \left(\frac{1}{3}\right)^{[\delta/\bar{\Delta}]}.$$

Moreover, if $\Delta \in \mathcal{P}_\sigma(a, b)$, then

$$(5.9) \quad \|Ds\|_{L_\infty[\alpha,\beta]} \le \frac{45}{2\bar{\Delta}} \sigma M \left(\frac{1}{3}\right)^{[\delta/\bar{\Delta}]},$$

and

$$(5.10) \quad \|D^2 s\|_{L_\infty[\alpha,\beta]} \le \frac{45}{(\bar{\Delta})^2} \sigma^2 M \left(\frac{1}{3}\right)^{[\delta/\bar{\Delta}]}.$$

Proof. The values s_i at the interior knots x_i, $1 \le i \le N-1$, are governed by the (N-1) linear equations of (2.3), i.e.,

$$a_i s_{i-1} + 3s_i + c_i s_{i+1} = 4a_i s_{i-1/2} + 4c_i s_{i+1/2}, \quad 1 \le i \le N-1.$$

Equivalently, in matrix notation, these equations can be expressed as

$$A\underline{s} = \underline{k},$$

where $\underline{s} \equiv (s_1, s_2, \cdots, s_{N-1})^T$, and where $\underline{k} \equiv (k_1, k_2, \cdots, k_{N-1})^T$, with $k_1 = 4a_1 s_{1/2} + 4c_1 s_{3/2} - a_1 s_0$, $k_i = 4a_i s_{i-1/2} + 4c_i s_{i+1/2}$, $1 < i < N-1$, and $k_{N-1} = 4a_{N-1} s_{N-3/2} + 4c_{N-1} s_{N-1/2} - c_{N-1} s_N$. As previously noted, A is a strictly diagonally dominant matrix, and is hence nonsingular. Thus, $\underline{s} = A^{-1}\underline{k}$, and on writing $A^{-1} \equiv (b_{i,j})$, then

$$(5.11) \quad s_i = \sum_{j=1}^{N-1} b_{i,j} k_j, \quad 1 \le i \le N-1.$$

Because of the explicit tridiagonal form of A, a result of Kershaw [6] gives us that

(5.12) $|b_{i,j}| \le \frac{3}{8}\left(\frac{1}{3}\right)^{|i-j|}, \quad 1 \le i,\ j \le N-1.$

Next, the assumption that $2\bar{\Delta} \le \delta$ allows us to deduce that

$$a' \le x_{i_0-1/2} < x_{i_0+1/2} < \cdots < x_{i_0+r+1/2} \le b'$$

with $i_0 > 1$, $r \ge 1$, and $i_0 + r + 1/2 < N-1$. Because of the hypothesis of (5.7i), it follows that $k_j = 0$, for all $i_0 \le j \le i_0 + r$. Moreover, $|k_i| \le 5M$ for all $1 \le i \le N-1$ from (5.7ii) and (2.4). Hence, with the bound of (5.12), we have from (5.11) that

$$|s_i| \le \frac{15M}{8}\left\{\sum_{i<i_0} \frac{1}{3^{|i-j|}} + \sum_{i>i_0+r} \frac{1}{3^{|i-j|}}\right\}, \quad 1 \le i \le N-1.$$

In particular, consider any $x_i \in [\alpha - \bar{\Delta}, \beta + \bar{\Delta}]$. On summing the above geometric series, it readily follows that

$$|s_i| \le \frac{45M}{8} \frac{1}{3^{[\delta/\bar{\Delta}]}} \quad \text{for any } x_i \in [\alpha - \bar{\Delta}, \beta + \bar{\Delta}].$$

Finally, using an argument similar to that used in the end of the proof of Theorem 3.1 shows that

$$\|s\|_{L_\infty[\alpha,\beta]} \le \frac{45M}{8} \cdot \frac{1}{3^{[\delta/\bar{\Delta}]}},$$

which establishes (5.8).

Assume now that $\Delta \in \mathcal{P}_\sigma(a, b)$. On each subinterval $[x_{i-1}, x_i]$, $1 \le i \le N$, the derivative of s is evidently linear, and therefore takes on its extreme values at the endpoints. By direct computation, it can be shown that

$$\|Ds\|_{L_\infty[x_{i-1},x_i]} = \frac{1}{(x_i - x_{i-1})} \max \left\{ |3s_{i-1} - 4s_{i-1/2} + s_i|; \right.$$

$$\left. |s_{i-1} - 4s_{i-1/2} + 3s_i| \right\}.$$

Thus, with the inequality of (5.8), this gives

$$\|Ds\|_{L_\infty[\alpha,\beta]} \le \frac{45}{2\bar{\Delta}} M\,\sigma\,\left(\frac{1}{3}\right)^{[\delta/\bar{\Delta}]},$$

the desired result of (5.9). As the second derivative of s is piecewise constant on (x_{i-1}, x_i), direct computation shows that

$$D^2 s(x) = \frac{4(s_{i-1} - 2s_{i-1/2} + s_i)}{(x_i - x_{i-1})^2}, \quad x \in (x_{i-1}, x_i).$$

Again, using the inequality of (5.8) gives the result of (5.10):

$$\|D^2 s\|_{L_\infty[\alpha,\beta]} \le \frac{45M\sigma^2}{(\bar{\Delta})^2} \left(\frac{1}{3}\right)^{[\delta/\bar{\Delta}]}. \qquad \text{Q.E.D.}$$

The local interpolating error bounds given in Theorem 5.3 for $\|D^j(f - P_\Delta f)\|_{L_\infty[\alpha,\beta]}$, $1 \le j \le 2$, requires the assumption that $\Delta \in \mathcal{P}_\sigma(a, b)$, and as such are weaker than the corresponding result of de Boor and Fix [2] (cf. (4.2)). However, because of the existence of Nord's well-known counterexample [13] for cubic spline interpolation, this assumption, $\Delta \in \mathcal{P}_\sigma(a, b)$, seems to be unavoidable.

For partitions $\Delta \in \mathcal{P}[a, b]$ which are uniform on $[a', b']$, that is, $h = x_i - x_{i-1}$ whenever both x_i and x_{i-1} are in $[a', b']$, the above local error bounds can be improved by a factor of h at various points of $[\alpha, \beta]$. Similar high order pointwise interpolation error bounds were obtained globally for quadratic splines in Marsden [11], and for cubic spline interpolants by Lucas [8].

Theorem 5.5. Given any $f \in C^{-1}[a, b]$ with $f \in C^4[a', b']$, and given any partition $\Delta \in \mathcal{P}[a, b]$ which is uniform of size h on $[a', b']$ and satisfies $2\overline{\Delta} \leq \delta$, the quadratic spline interpolant s of f in $Sp(2, \Delta)$ in the sense of (3.1) satisfies

$$(5.13)\quad \begin{cases} \text{i)}\ |f(x_i) - s(x_i)| \leq Kh^4 & \text{for } x_i \in [\alpha,\beta], \\ \text{ii)}\ |D(f(x_i+\lambda h) - s(x_i+\lambda h))| \leq Kh^3 & \text{for } x_i+\lambda h \in [\alpha,\beta], \\ \text{iii)}\ |D^2(f(x_i+\tfrac{1}{2}h) - s(x_i+\tfrac{1}{2}h))| \leq Kh^2 & \text{for } x_i+\tfrac{1}{2}h \in [\alpha,\beta], \end{cases}$$

where K is a constant independent of f and Δ, $\lambda \equiv (3 \pm \sqrt{3})/6$.

Proof. From (3.2), the errors $e_i \equiv f_i - s_i$ at the interior knots x_i, $1 \leq i \leq N-1$, are seen to satisfy the N-1 linear equations

$$(5.14)\quad a_i e_{i-1} + 3e_i + c_i e_{i+1} = a_i f_{i-1} - 4a_i f_{i-1/2} + 3f_i - 4c_i f_{i+1/2} + c_i f_{i+1},$$

for $1 \leq i \leq N-1$. Note that the associated coefficient matrix for the e_i's, determined from the left-hand side of (5.14), is identical to the matrix A defined in the proof of Lemma 5.4. Furthermore, because Δ is uniform on $[a', b']$ and $f \in C^4[a', b']$, Taylor series expansions of f about any $x_i \in [a' + \bar{\Delta}, b' - \bar{\Delta}]$ show that the corresponding equations of (5.14) reduce to

$$(5.15)\quad \frac{1}{2}e_{i-1} + 3e_i + \frac{1}{2}e_{i-1} = \frac{1}{2}f_{i-1} - 2f_{i-1/2} + 3f_i - 2f_{i+1/2} + \frac{1}{2}f_{i+1}$$

$$= O(h^4), \quad 1 \le i \le N-1.$$

Inequality (5.13i) can now be obtained by utilizing, as in the proof of Lemma 5.4, the exponential damping of the off-diagonal entries of A^{-1}.

To establish (5.13ii), set $x = x_i + \lambda h$, for $x_i \in [a, b - \bar{\Delta}]$ and $0 \le \lambda \le 1$. Then, by direct computations

$$Ds(x_i + \lambda h) = \frac{2}{h}\{(2\lambda - \frac{3}{2})s_i - 2(2\lambda - 1)s_{i+1/2} + (2\lambda - \frac{1}{2})s_{i+1}\}$$

and from Taylor series expansions about $x = x_i + \lambda h$,

$$D(f(x_i + \lambda h) - s(x_i + \lambda h)) = \frac{2}{h}\{(2\lambda - \frac{3}{2})e_i - 2(2\lambda - 1)e_{i+1/2} +$$

$$(2\lambda - 1/2)e_{i+1}\} + \frac{h^2}{3!}\{3\lambda^2 - 3\lambda + \frac{1}{2}\}D^3f(x_i + \lambda h) + O(h^3).$$

But $3\lambda^2 - 3\lambda + \frac{1}{2} = 0$ if and only if $\lambda = (3 \pm \sqrt{3})/6$. Assigning either of these values of λ and making use of (5.13i) gives the desired result (5.13ii). Inequality (5.13iii) follows readily from the identity

$$D^2f_{i+1/2} - D^2s_{i+1/2} = 4h^{-2}\{e_{i+1} - 2e_{i+1/2} + e_i\} - \frac{h^2}{48}f^{(4)}(\xi),$$

for some $\xi \in (x_i, x_{i+1})$, $x_i, x_{i+1} \in [a', b']$. Q.E.D.

In order to illustrate numerically some of the results of this section, we interpolate the function $f \in C^{-1}[0, 1]$, defined by

$$(5.16) \quad f(x) = \begin{cases} \sin 2\pi x, & 0 \le x \le .5, \\ -1, & .5 < x \le 1, \end{cases}$$

by its unique interpolant, $P_\Delta f$, in $Sp(2, \Delta)$, using uniform partitions of [0, 1]. Column 2 of Table 5.1 contains the maximum interpolation error over the subinterval [0, .25]. Assuming that $\|f - P_\Delta f\|_{L_\infty[0, .25]}$ behaves, as a function of $\bar{\Delta}$, like $C(\bar{\Delta})^\beta$, then one can estimate the exponent β by successively computing

$$\beta = \ln\left\{\frac{\|f - P_{\Delta_1} f\|_{L_\infty[0,.25]}}{\|f - P_{\Delta_2} f\|_{L_\infty[0,.25]}}\right\} \Big/ \ln(\bar{\Delta}_1/\bar{\Delta}_2).$$

These computations are contained in column 3 of Table 5.1. Note that Theorem 5.1 gives us that this exponent β theoretically tends to 3, as $\bar{\Delta} \to 0$. Column 4 of Table 5.1 contains the interpolation error at $x = .25$, a common nodal point, and column 5 gives the associated observed exponent of $\bar{\Delta}$ for the numbers in column 4. Note that Theorem 5.5 shows that these numbers β in column 5 theoretically tend to 4, as $\bar{\Delta} \to 0$.

Table 5.1

$\bar{\Delta}$	$\|f-P_{\Delta}f\|_{L_\infty[0,.25]}$	β	$\|f(.25)-(P_{\Delta}f)(.25)\|$	β
1/16	$.561 \cdot 10^{-3}$	--	$.561 \cdot 10^{-3}$	--
1/32	$.614 \cdot 10^{-4}$	3.19	$.120 \cdot 10^{-4}$	5.55
1/48	$.182 \cdot 10^{-4}$	3.00	$.230 \cdot 10^{-5}$	4.07
1/64	$.778 \cdot 10^{-5}$	3.00	$.726 \cdot 10^{-6}$	4.01

References

1. Garrett Birkhoff, and Carl de Boor, "Error bounds for spline interpolation," J. Math. Mech. 13 (1964), 827-835.

2. C. de Boor and G. J. Fix, "Spline approximation by quasi-interpolants," J. Approx. Theory, 8 (1973), 19-45.

3. E. W. Cheney and F. Schurer, "Convergence of cubic spline interpolants," J. Approx. Theory 3 (1970), 114-116.

4. C. A. Hall, "Uniform convergence of cubic spline interpolants," J. Approx. Theory 7 (1973), 71-75.

5. W. J. Kammerer and G. W. Reddien, "Local convergence of smooth cubic spline interpolation," SIAM J. Numer. Anal. 9 (1972), 687-694.

6. D. Kershaw, "Inequalities on the elements of the inverse of a certain tri-diagonal matrix," Math. Comp. 24 (1970), 155-158.

7. G. G. Lorentz, Approximation of Functions. New York: Holt, Rinehart, and Winston, 1966.

8. T. R. Lucas, "Error bounds for interpolating cubic splines under various end conditions," SIAM J. Numer. Anal. (to appear).

9. Tom Lyche and Larry L. Schumaker, "On the convergence of cubic interpolating splines," Spline Functions and Approximation Theory (A. Meir and A. Sharma, ed.), Birkhäuser Verlag, Basel, 1973, 169-189.

10. Martin Marsden, "Cubic spline interpolation of continuous functions," J. Approx. Theory (to appear).

11. M. J. Marsden, "Quadratic spline interpolation," to appear in Bull. Amer. Math. Soc.

12. A. Meir and A. Sharma, "On uniform approximation by cubic splines," J. Approx. Theory 2 (1969), 270-274.

13. S. Nord, "Approximation properties of the spline fit," BIT 7 (1967), 132-144.

14. B. K. Swartz and R. S. Varga, "Error bounds for spline and L-spline interpolation," J. Approx. Theory 6 (1972), 6-49.

15. Richard S. Varga, Matrix Iterative Analysis. Englewood Cliffs, New York: Prentice-Hall, 1962.

SOME LINEAR AND NON-LINEAR PROBLEMS IN FLUID MECHANICS FEM FORMULATION

O.C. Zienkiewicz

University College of Swansea, U.K.

Foreword

To illustrate the discretisation process of the finite element method in non-structural situations this lecture chooses to concentrate on the subject of the flow of real fluids. Within this one area of activity we can illustrate the use of virtual work concepts, derivation of various linear and non linear discretised forms and for the special case of irrotational flow, the specialisation to a simple LAPLACE equation.

The lecture is intended to be directly instructive and does not attempt to survey the contributions made to the literature of the subject referring only to selected texts or papers where illustrative examples can be found. A complete "state of the art" survey of flow problems will be presented at an International Conference on Finite Elements in Flow Problems to be held in January 1974 at Swansea and the interested reader is referred to the proceedings of that conference.

1. INTRODUCTION - BASIC FORMULATION FOR VISCOUS FLOW

By contrast to the solid mechanics problem where we were primarily concerned with static response and concentrated often the main interest in a displacement of a material point - in fluid mechanics the main variable of interest is the velocity at a point in space. The Lagrangian description was predominant in solids while here we shall follow an Eulerian one. Nevertheless the similarity of general concepts is great and we shall therefore borrow heavily on the methodology used in solids (1).

Let u, v, w describe the three, Cartesian components of the velocity $\underline{u}$ of a point x,y,z. Further let $\underline{b}$ denote the body forces per unit volume in part due to external causes ($\underline{b}_o$) and in part representing the dynamic acceleration effects. Thus,

$$\underline{b} = \underline{b}_o - \rho\underline{a} \qquad 1$$

in which $\underline{a}$ stands for the acceleration of a point in the fluid.

If equilibrium is considered between the internal stresses $\underline{\sigma}$ and body forces we have precisely the same equation set as in solid mechanics problem i.e.

$$\underline{L}_E\, \underline{\sigma} + \underline{b} = 0 \qquad 2$$

or explicitly a set

$$\frac{\partial}{\partial x}\sigma_x + \frac{\partial}{\partial y}\sigma_{xy} + \frac{\partial}{\partial z}\sigma_{xz} + X = 0 \qquad 3$$

with X being the x component of $\underline{b}$ etc.

A major difference from solid mechanics is due however to the Eulerian description of motion which even in steady state cases results in an acceleration $\underline{a}$.

Consider the acceleration component in direction x for a mass point which has a velocity $\underline{u}$ at a point of space xyz. Its velocity rate of change for a particle of fluid depends not only on rates of change of $\underline{u}$ with respect to time but also on the changes of position. Thus

$$a_x = \frac{D}{Dt}u = \frac{\partial}{\partial t}u + \frac{\partial u}{\partial x}\cdot\frac{dx}{dt} + \frac{\partial u}{\partial y}\cdot\frac{dy}{dt} + \frac{\partial u}{\partial z}\cdot\frac{dz}{dt} \qquad 4$$

As $\frac{dx}{dt} = u$ etc.

we have a definition of the 'convective' acceleration operator

$$\frac{D}{dt} \equiv \frac{\partial}{\partial t} + u\frac{\partial}{\partial x} + v\frac{\partial}{\partial y} + w\frac{\partial}{\partial z} = \frac{\partial}{\partial t} + \underline{u}^T\cdot \text{grad}\,\underline{u} \qquad 5$$

Here lies the major difference from solid mechanics where, as the displacements were referred to a particle and not to an element of space only the simple differentiation of displacement with respect to time sufficed.

To complete the formulation of a solid mechanics problem we introduced a definition of strain in terms of displacements and a constitutive law defining a stress-strain relation (1). For fluids we shall proceed similarly. First a rate of deformation, $\underline{\dot{\varepsilon}}$, is defined in terms of velocities. Thus

$$\underline{\dot{\varepsilon}} = \underline{L}\ \underline{u} \qquad 6$$

with $\dot{\varepsilon}_{xx} = \frac{\partial u}{\partial x}$ etc. defining the operator L.

If tensorial representation is preferred the equivalent definition is

$$\dot{\varepsilon}_{ij} = \frac{1}{2}\left(\frac{\partial u_i}{\partial x_j} + \frac{\partial u_j}{\partial x_i}\right) \qquad i = 1 - 3 \qquad 6a$$

The constitutive relationship for fluids is more complex than in the solid mechanics problems, as in general stresses depend not only on rates of strain but also on the strain itself. We shall therefore restrict our attention here to incompressible flow for which the rate of volumetric straining is zero, i.e.

$$\varepsilon_v = \varepsilon_{xx} + \varepsilon_{yy} + \varepsilon_{zz} \equiv \varepsilon_{ii} = \text{div}\ \underline{u} = 0 \qquad 7$$

For such fluids the mean stress σ is not defined and has to be sought from equilibrium relations. Defining the mean stress σ on the pressure p as

$$\sigma = \frac{\sigma_{xx} + \sigma_{yy} + \sigma_{zz}}{3} = -\ p \qquad 8$$

we can write the deviatoric portion of the stress $\underline{S}$ as a function of strain rate $\underline{\dot{\varepsilon}}$ in a matrix notation as

$$\underline{S} = \underline{\sigma} - \begin{Bmatrix} 1 \\ 1 \\ 1 \\ 0 \\ 0 \\ 0 \end{Bmatrix} \sigma \overline{\overline{\ }} \underline{\sigma} + \underline{Mp} \equiv \underline{R\sigma} = \underline{f}(\underline{\dot{\varepsilon}})$$

with

$$\underline{S} = \underline{D}\ \underline{\dot{\varepsilon}}\ ;\ \underline{R} = \begin{bmatrix} \frac{2}{3} & -\frac{1}{3} & -\frac{1}{3} & 0 & 0 & 0 \\ & \frac{2}{3} & -\frac{1}{3} & 0 & 0 & 0 \\ & & \frac{2}{3} & 0 & 0 & 0 \\ \text{SYM} & & & 1 & 0 & 0 \\ & & & & 1 & 0 \\ & & & & & 1 \end{bmatrix} \qquad 9$$

In a linear fluid with $\underline{D}$ being a matrix of constants. In tensorial equivalent we can rewrite above as

$$S_{ij} = \sigma_{ij} - \frac{1}{3}\sigma_{ii} = D_{ijk\ell}\,\dot{\varepsilon}_{k\ell} \qquad 9a$$

Above is entirely analogous to the definition of behaviour of elastic solids which are incompressible with $\underline{D}$ playing the role of the matrix or elastic constants. For isotropic linear behaviour it can be readily shown that we can write

$$\underline{D} = \mu \begin{vmatrix} 2 & & & & & 0 \\ & 2 & & & & \\ & & 2 & & & \\ & & & 1 & & \\ & & & & 1 & \\ 0 & & & & & 1 \end{vmatrix} = \mu\underline{D}^{o} \quad \text{OR} \quad D_{ijk\ell} = 2\mu \qquad 10$$

in which μ is known as viscosity and D^{o} a diagonal matrix. In general μ will be a function of $\dot{\underline{\varepsilon}}$ and the formulation that follows is applicable in this form to Non-Newtonian (non-linear) fluids.

The viscous flow problem is now fully defined and we can formulate its approximate solution mathematically proceeding formally by Galerkin or other weighting method. However an examination of the equations governing the flow and of the boundary conditions - which specify either tractions or velocities at all external boundaries permits us to adapt here the procedures used in solid mechanics. In particular the 'virtual work principle' can be applied with virtual velocities playing the part of virtual displacements used in solid mechanics.

Indeed one can conclude that the solution of a viscous flow problem is identical to the solution of an equivalent incompressible elastic problem in which displacements are replaced by velocities and the body forces described by equation 1 are inserted.

Thus all the techniques available in the literature for the solution of one problem are available to the other - applicability being direct if the fluid flow is sufficiently slow so that acceleration effects can be disregarded.

2. VISCOUS FLOW - VELOCITY FORMULATION

2.1 Virtual Work Statements

The equilibrium statement 3 can be replaced by an equivalent virtual work statement requiring that for any virtual velocity and strain rate changes $\delta\underline{u}$ and $\delta\dot{\varepsilon}$ which are compatible, external and internal rates of work are identical. Thus we can write

$$\int_{\Omega} \delta\dot{\underline{\varepsilon}}^T \underline{\sigma} \, d\Omega - \int_{\Omega} \delta\underline{u}^T \underline{b} \, d\Omega - \int_{\Gamma_\sigma} \delta\underline{u}^T \bar{t} \, d\Gamma = 0 \qquad 11$$

for any flow domain Ω in which tractions are specified on boundary Γ_σ and where $\delta\underline{u}$ is zero on boundary Γu where velocities are given. Further for compatibility

$$\delta\dot{\underline{\varepsilon}} = \underline{L} \, \delta \, \underline{u} \quad \text{and} \quad \delta\underline{u} = 0 \text{ on } \Gamma_u \qquad 12$$

As σ is not uniquely defined by $\dot{\varepsilon}$ (being indeterminate due to the undefined pressure p, an additional equation needs to be written to enforce the incompressibility. As for any pressure variation δp internal work is zero due to incompressibility we can write

$$\int_{\Omega} \delta p \, \varepsilon_v \, d\Omega = 0 \qquad 13$$

Inserting the constitutive relation 9 into 11 we can rewrite this as

$$\int_{\Omega} \delta\dot{\underline{\varepsilon}}^T \mu \underline{D} \dot{\varepsilon} \, d\Omega + \int_{\Omega} \delta \, \varepsilon_v \, p \, d\Omega - \int_{\Omega} \delta\underline{u}^T \underline{b} \, d\Omega$$

$$- \int_{\Gamma_\sigma} \delta u^T \bar{t} \, d\Gamma = 0 \qquad 14$$

Observing that by 9

$$\varepsilon_v = \varepsilon_{xx} + \varepsilon_{yy} + \varepsilon_{zz} \equiv \underline{M}^T \varepsilon \qquad 15$$

Equations 13 and 14 lead directly to a discretisation. We have now several choices open: First we can allow an unrestricted definition of the fields of $\underline{u}$ and $\underline{p}$ and proceed using both equations. Second we can confine our attention to velocity fields which automatically satisfy incompressibility. In the latter case the equation 13 as well as the second term of equation 14 disappear and the full approximation involves only velocity parameters. We shall explore both formulations in turn.

The reader could at this stage with profit rederive the approximation equations using a straightforward Galerkin procedure - and as usual will observe the integration and interpretation difficulties present.

2.2 Discretisation with velocity and pressure fields

In the usual manner we describe the displacement and pressure fields by trial functions as

$$\underline{u} = \sum N_i^u a_i^u = \underline{N}^u \underline{a}^u$$

$$\underline{p} = \sum N_i^p a_i^p = \underline{N}^p \underline{a}^p \qquad 16$$

in which $\underline{N}^u$ and $\underline{N}^p$ are appropriate shape functions, defined element by element. It will be observed from the nature of integrals involved that we require C^o continuity for the velocity field but discontinuous functions can be used to describe the pressure field.

Writing $\delta\underline{\varepsilon} = \underline{L}\,\underline{u} = (\underline{L}\,\underline{N}^u)\,\delta\underline{a}^u; \quad \delta\varepsilon_v = M\delta\underline{\varepsilon}$

$$\delta p = \underline{N}^p\,\delta\,\underline{u}^p \qquad 17$$

and observing that 13 and 14 are true for all variations we have from 14

$$\left(\int_\Omega (\underline{L}\,\underline{N}^u)^T \mu\,\underline{D}^o\,(\underline{L}\,\underline{N}^u)\,d\Omega\right)\underline{a}^u + \left(\int_\Omega (\underline{M}^T\,\underline{L}\,\underline{N}^u)^T\,\underline{N}^p\,d\Omega)\underline{a}^p\right)$$

$$- \int_\Omega \underline{N}^{u\,T}\,\underline{b}\,d\Omega - \int_{\Gamma_\sigma} \underline{N}^{u\,T}\,\bar{t}\,d\Gamma = 0 \qquad 18$$

and from 13

$$\int_\Omega \underline{N}^{pT}\,\underline{M}^T\,\underline{L}\,\underline{N}^u\,d\Omega\,\underline{a}^u = 0 \qquad 19$$

For cases of slow steady state viscosity flow where $\underline{b} = \underline{b}_o$ this results in a simple symmetric set of equations which can be written as

$$\begin{bmatrix} \underline{K}^u & \underline{K}^p \\ \underline{K}^{pT} & 0 \end{bmatrix} \begin{Bmatrix} \underline{a}^u \\ \underline{a}^p \end{Bmatrix} + \begin{Bmatrix} f^u \\ 0 \end{Bmatrix} = 0 \qquad 20$$

where

$$\underline{K}_{ij}^u = \int_\Omega (\underline{L}\,N_i^u)^T\,\mu\,D_o\,(\underline{L}\,\underline{N}_j^u)\,d\Omega$$

$$\underline{K}_{ij}^p = \int_\Omega (\underline{M}^T\,\underline{L}\,\underline{N}_i^u)^T\,\underline{N}_j^p\,d\Omega \qquad 21$$

$$\underline{f}_i^u = \int_\Omega \underline{N}_i^{u\,T}\,\underline{b}\,d\Omega - \int_{\Gamma_\sigma} N^{u\,T}\,\bar{t}\,d\Gamma$$

Indeed this formulation is almost identical to that used in linear, incompressible elasticity and which has been derived by imposing constraints on an energy functional (2). If μ is velocity independent 20 represents a simple linear equation system but formulation is generally valid.

When considering unsteady state or when flow is not slow then in general coefficients $\underline{a}$ are time dependent and $\underline{f}^u$ depends on these even in steady state due to the convective terms.

Returning to eq. 1 and the explicit expression for the acceleration we note that

$$\underline{b} = \underline{b}_o + \rho \frac{\partial u}{\partial t} + \rho \, \underline{J} \, \underline{u} \qquad 22$$

where

$$\underline{J} = \begin{bmatrix} \frac{\partial u}{\partial x} & \frac{\partial u}{\partial y} & \frac{\partial u}{\partial z} \\ \frac{\partial v}{\partial x} & \frac{\partial v}{\partial y} & \frac{\partial v}{\partial z} \\ \frac{\partial w}{\partial x} & \frac{\partial w}{\partial y} & \frac{\partial w}{\partial z} \end{bmatrix}$$

The term $\underline{f}^u$ of equation 20 now has in addition to the contribution given by constant body forces and boundary traction, f_o, a form

$$\bar{\underline{f}}^u = \left(\int_\Omega \underline{N}^{uT} \rho \, \underline{N}^u \, d\Omega\right) \frac{d\underline{a}^u}{dt} + \left(\int \underline{N}^{uT} \, \underline{J} \, \underline{N}^u \, d\Omega\right) \underline{a}^u$$

$$= M^u \frac{da^u}{dt} + \bar{\underline{K}}^u \, \underline{a}^u \qquad 23$$

The discretised equation 20 becomes now

$$\begin{bmatrix} (\underline{K}^u + \bar{\underline{K}}^u) & \underline{K}^p \\ \underline{K}^{pT} & 0 \end{bmatrix} \begin{Bmatrix} \underline{a}^u \\ \underline{a}^p \end{Bmatrix} + \begin{vmatrix} M^u & 0 \\ 0 & 0 \end{vmatrix} \frac{d}{dt} \begin{Bmatrix} \underline{a}^u \\ \underline{a}^p \end{Bmatrix} + \begin{Bmatrix} f_o \\ 0 \end{Bmatrix} = 0 \qquad 24$$

In above the transient can be solved by time stepping procedures but even in steady state the form matrix equations obtained by omitting the second term is non-linear and non-symmetric.

The matrix $\bar{\underline{K}}^u$ depends on the current velocity and its form is non symmetric. This presents difficulties in solving viscous flow problems with appreciable inertia and several alternative procedures have been used (3,4). In some the non-linear and non-symmetric matrix $\bar{\underline{K}}^u$ is taken care of by repeated iteration using only the symmetric part - in others attempts at a direct solution of the non-linear equations have been made using a non-symmetric solution scheme.

2.3 Discretisation using incompressible velocity fields

The most usual procedure of describing incompressible velocities is by the use of stream function in two dimensional problems or by introduction of a vector potential in three dimensions.

Thus if we confine our attention to plane flow with u and v velocity components in x and y directions we can write

$$u = -\frac{\partial\psi}{\partial y} \qquad v = \frac{\partial\psi}{\partial x} \qquad \text{OR} \qquad \underline{u} = \hat{\underline{L}}\ \psi \tag{25}$$

It is easily verified that

$$\varepsilon_v = \varepsilon_{xx} + \varepsilon_{yy} = \frac{\partial u}{\partial x} + \frac{\partial v}{\partial y} \equiv 0 \tag{26}$$

In an axisymmetric flow similarly we can write for radial and axial velocity components

$$u = -\frac{1}{r}\frac{\partial\psi}{\partial y} \qquad v = \frac{1}{r}\frac{\partial\psi}{\partial x} \tag{27}$$

and once again incompressibility is obtained.

Finally in three dimensional flow we define the velocity in terms of a vector potential with three components

$$\underline{\psi}^T = [\psi_x, \psi_y, \psi_z] \tag{28}$$

as

$$\underline{u} = \text{curl}\ \underline{A} = \hat{L}\ \psi \tag{29}$$

where

$$\hat{\underline{L}} = \begin{bmatrix} \frac{\partial}{\partial z} & 0 & -\frac{\partial}{\partial y} \\ 0 & \frac{\partial}{\partial z} & -\frac{\partial}{\partial x} \\ \frac{\partial}{\partial y} & -\frac{\partial}{\partial x} & 0 \end{bmatrix}$$

Again it is easily verified that incompressibility is satisfied as

$$\varepsilon_v = \frac{\partial u}{\partial x} + \frac{\partial v}{\partial y} + \frac{\partial w}{\partial z} \equiv 0 \tag{30}$$

(as ε_v = div $\underline{u}$ and div curl $\underline{\psi} \equiv 0$).

With velocities specified on the boundaries it is possible to determine the stream function and its normal gradient (or vector potential components) there to within an arbitrary constant. For discretisation therefore we can assume an expansion for ψ

$$\psi = \sum \underline{N}_i \underline{a}_i = \underline{N}\,\underline{a} \qquad 31$$

and use the virtual work expression 14 with the second term dropped, (as it now is identically zero) i.e.

$$\int_\Omega \delta\underline{\dot{\varepsilon}}\,\mu\,\underline{D}_o\,\dot{\varepsilon}\,d\Omega - \int_\Omega \delta\,\underline{u}^T\,\underline{b}\,d\Omega - \int_{\Gamma_\sigma} \delta u^T\,\bar{t}\,d\Gamma = 0 \qquad 32$$

Writing for all cases considered above

$$\underline{u} = \underline{\hat{L}}\,\psi = \underline{\hat{L}}\,\underline{N}\,\underline{a}\ ; \quad \dot{\varepsilon} = \underline{L}\,\underline{u} = \underline{L}\,\underline{\hat{L}}\,\underline{N}\,\underline{a} \qquad 33$$

with corresponding variation we can immediately obtain the discretised form of equations for which $\underline{a}$ can be obtained as (as usual taking $\underline{a}^T$ outside in 32 after substitution and equating the multiplier to zero)

$$\left(\int_\Omega (\underline{L}\,\underline{\hat{L}}\,\underline{N})^T\,\mu\,\underline{D}_o\,(\underline{L}\,\underline{\hat{L}}\,\underline{N})\,d\Omega\right)\underline{a} - \int_\Omega (\underline{\hat{L}}\,N)^T\underline{b}\,d\Omega$$

$$- \int_{\Gamma_\sigma} (\underline{\hat{L}}\,\underline{N})^T\,\bar{t}\,d\Gamma = 0$$

or the usual form

$$\underline{K}\,\underline{a} - f = 0 \qquad 35$$

with expressions for K_{ij} and f_i apparent from 34.

Immediately an observation can be made that the shape functions $\underline{N}$ now require C' continuity as second derivatives operate in these in the integrals. While in two dimensional problems the use of such functions presents little difficulty in three dimensions no satisfactory piecewise defined functions are available.

Confining our attention to the scalar stream function ψ defined for axisymmetric or plane problems it is readily seen that the same shape functions as used for plate bending analysis are available. It is therefore possible to use any of the numerous plate functions for solution of viscous flow.

Indeed for the linear case of slow viscous flow equations and the whole formulation may be identified with plate bending equation and any standard plate bending program adapted immediately.

While in the first type of formulation section 2.2 we have 'borrowed' heavily from previous methods used extensively in the solution of incompressible solids - in the second type (section 2.3) we have introduced standard fluid mechanics concepts for enforcing incompressibility. These appear not to have been used widely in solid mechanics and a "reverse borrowing" is obviously possible. The stream function concept can be used directly in solid mechanics and only recently has such a development been put into practice (5).

Once again inclusion of the dynamic term can be made pursuing the process of modifying the $\underline{f}$ term - as described in the previous section.

3. VISCOUS FLOW - EQUILIBRIUM AND MIXED FORMULATION

The need for enforcing incompressibility has presented some difficulties in the velocity type of formulation used in the previous section this arising because the stresses are not completely defined by the strain rates (vide eq. 9). On the other hand stresses define uniquely the strain rates and it is obviously possible to use the equivalents of equilibrium virtual work statements or of the "mixed" formulations well known in solid mechanics with advantage. Possibilities here are enormous and have only been barely explored. We shall here restrict ourselves to a brief statement of the equilibrium formulation.

If the unknown function is the stress field σ which is so chosen as to satisfy exactly the equilibrium conditions then the virtual work done by 'compatible' strain rates yields

$$\int_{\Omega} \delta \underline{\sigma}^T \dot{\varepsilon} \, d\Omega = \int_{\Gamma_u} \delta \underline{t}^T \bar{\underline{u}} \, d\Gamma \qquad 36$$

in which Γ_u is the portion of the boundary on which velocities $u = \bar{u}$ are specified, and δt are the boundary tractions resulting from stresses $\delta\underline{\sigma}$

$$\delta\underline{t} = \underline{G} \, \delta \, \underline{\sigma} \qquad 37$$

where $\underline{G}$ is a suitable matrix of direction cosines of the normal to the surface. From eq. 9 the strain rates are defined in terms of stresses as

$$\underline{R} \, \underline{\sigma} = \mu \, \underline{D}_o \, \dot{\underline{\varepsilon}} \qquad 38$$

or

$$\dot{\underline{\varepsilon}} = \frac{1}{\mu} (\underline{D}_o^{-1} \, R) \, \underline{\sigma} = \frac{1}{\mu} \underline{C}_o \, \underline{\sigma} \qquad 39$$

in which

$$\underset{\sim}{C_o} = \begin{bmatrix} \frac{1}{3} & -\frac{1}{6} & -\frac{1}{6} & 0 & 0 & 0 \\ & \frac{1}{3} & -\frac{1}{6} & 0 & 0 & 0 \\ & & \frac{1}{3} & 0 & 0 & 0 \\ \text{SYM} & & & 1 & 0 & 0 \\ & & & & 1 & 0 \\ & & & & & 1 \end{bmatrix}$$

To achieve an equilibrating field stresses can be defined in terms of a stress function set $\underline{\Phi}$

$$\underline{\sigma} = \underline{E}\ \Phi + \underline{\Omega} \qquad 40$$

where $\underline{\Omega}$ is a particular solution which equibrates the applied body forces and $\underline{\Phi}$ is so constrained as to satisfy prescribed boundary tractions.

Use of stress function in context of solid mechanics has been pioneered by Veubeke and Zienkiewicz (6) with a subsequent development by Sander (7) which in a sense is an application of certain special mixed formulations. Details of the procedure are discussed elsewhere but some of the difficulties in the fluid mechanics context should now be noted.

First if the flow is not so slow that dynamic terms can be neglected these will appear in the inhomogeneous terms defining the stress field $\underline{\Omega}$ in terms of velocities. As velocities can only be obtained by integration of the stress field similar difficulties will arise at those encountered in dynamics of solids when equilibrating forms are used. These difficulties can be overcome as shown by Tabarrock (8) but for practical application the equilibrating formulation appears only simple in cases of slow viscous flow.

Further in three dimensional problems the use of the stress function necessitates again C' continuity to be introduced in appropriate stress functions with near impossibility of achieving this in practice. Two dimensional use of the Airy stress function is however practicable and useful.

Possibilities of using a "mixed" formulation in which both stress and velocity field simultaneously appear has not yet been explored and presents a fruitful field for research.

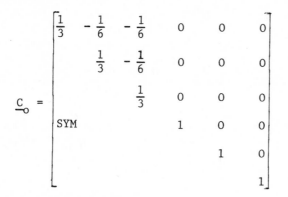

4 SOME OTHER SOLUTION POSSIBILITIES

In preceding we have used simple virtual work or which is equivalent Galerkin formulation. (Virtual work avoiding the difficulties of integration by parts and giving a direct physical interpretation of the various terms.) Other forms of discretisation are obviously possible and have been used in practice.

In some, following classical procedures of fluid dynamics, the governing equations are rewritten in terms of both stream function and vorticity. Direct approximation can then be used with Galerkin or other processes. Alternatives with the use of least square principle are possible - and can be applied directly to the equations in terms of all the variables.

A simple direct possibility can be used if we consider the analogy of fluid flow and incompressible elastic formulation. In the latter, to sidestep the difficulties involved due to incompressibility in a direct displacement formulation, the engineers have often used standard displacement programs with near incompressibility imposed. In terms of elastic constants this is equivalent to using a high Poissons ratio (say 0.49) in place of 0.5 for which a singularity arises. For simple finite element representation this usually leads to inaccurate results but with isoparametric parabolic elements good accuracy can be found with Poissons ratio as high as 0.4999 - especially if "reduced integration" of 2x2 Gauss points is used.

This procedure is particularly simple as existing finite element programs can be used. Elsewhere (9) it was shown that the approximation is equivalent to introduction of the incompressibility constraints via the use of "penalty functions".

5. SOME ILLUSTRATIVE APPLICATIONS

5.1 Creeping flow with constant viscosity

One of the first solutions to such problem was achieved by Atkinson et al (10) with a stream function formulation.

In this situation the velocities have been assumed entirely specified on the external boundary and no prescribed tractions introduced. The singly connected boundary allows the stream function and its normal gradient to be determined from the velocity definitions (eq. 25) and no problems arise (an arbitrary constant in the value of ψ is obtained by specifying this at some point of the boundary).

If the boundary is 'multiply connected' or when a singly connected boundary two or more separate regions exist where velocities are specified with tractions given between a difficulty in the use of stream function exists as the arbitrary value can only be specified at one point of the region and there the stream function is not uniquely defined.

For such problems auxiliary conditions have to be introduced involving the prescription of the rate of work being done on boundaries (11). Solution can not now be accomplished easily in one stage and for such problems the more direct discretisation of section 2.2 is preferable.

Numerous direct solutions in terms of velocity and pressure is given for some creeping flow problems in ref. 3.

5.2 Creeping flow with variable viscosity - Non-Newtonian flow

In Non-Newtonian fluids the viscosity μ depends in same manner as the rate of straining $\underline{\dot{\varepsilon}}$.

$$\mu = \mu(\underline{\dot{\varepsilon}}) \qquad 41$$

Here using the formulations of section 2.2 or of 2.3 we have (in the absence of dynamic terms) a discretised system of equations in the form

$$\underline{K}(\mu)\underline{a} - \underline{f} = 0 \qquad 42$$

The matrix $\underline{K}$ is dependent on $\underline{\dot{\varepsilon}}$ and hence on $\underline{a}$. A simple iterative procedure can now be adopted - and has been shown to converge quite rapidly even with substantially non linear behaviour.

Assuming some value of $\mu = \mu^o$ the first solution is obtained

$$\underline{a}^1 = \underline{K}_o^{-1} \underline{f} \qquad 43$$

and hence $\underline{\dot{\varepsilon}}$ and new value of μ at all points of the region is available to compute $\underline{K}_1$. The next approximation is

$$\underline{a}^2 = \underline{K}_1^{-1} \underline{f} \text{ etc.} \qquad 44$$

leading to a standard iterative algorithm

$$\underline{a}^n = \underline{K}_{n-1}^{-1} \underline{f} \qquad 45$$

It is usual to assume that viscosity is simply a function of the second strain rate invariant $\underline{\dot{\bar{\varepsilon}}}$

$$\dot{\bar{\varepsilon}} = \sqrt{2\, \dot{\varepsilon}_{ij}\, \dot{\varepsilon}_{ij}}$$

$$= 2\dot{\varepsilon}_{xx}{}^2 + 2\dot{\varepsilon}_{yy}{}^2 + 2\dot{\varepsilon}_{zz}{}^2 + (2\dot{\varepsilon}_{xy})^2 + (2\dot{\varepsilon}_{yz})^2 + (2\dot{\varepsilon}_{zx})^2 \qquad 46$$

A frequently used expression is of a form

$$\mu = \mu_o\ (\dot{\bar{\varepsilon}})^{n-1} \qquad 47$$

with index $n \leq 1$

Some solutions for a fluid of this type obtained by Palit and Fenner (12) using a stream function formulation and 'incompatible' triangular elements.

5.3 Plastic or viscoplastic behaviour of extruded metals - a case of Non-Newtonian flow

A particularly interesting case of non-Newtonian creeping flow is that of a Bingham fluid or its generalisation - the visco-plastic material.

Such materials behave as solids exhibiting a zero rate of straining for stresses which, are in some measure, below a threshhold or yield value. When this yield is exceeded flow begins at a rate which is a function of the excess stress.

Let $F(\underline{\sigma}) = 0$ represent this yield condition and we shall assume therefore this if $F(\underline{\sigma}) < 0$ no flow occurs. Assuming further that the various components of strain rate are proportional to the gradients of F with respect to these (i.e. the so called associated plasticity condition) we can describe with some generality the behaviour of the material by writing

$$\dot{\underline{\varepsilon}} = \frac{1}{2\bar{\mu}} \langle F^n \rangle \frac{\partial F}{\partial \underline{\sigma}} \qquad 48a$$

or in tensorial form

$$\dot{\varepsilon}_{ij} = \frac{1}{2\bar{\mu}} \langle F^n \rangle \frac{\partial F}{\partial \sigma_{ij}} \qquad 48b$$

where < > means that

$$\begin{aligned} \langle F \rangle &= 0 \quad \text{if} \quad F < 0 \\ \langle F \rangle &= F \quad \text{if} \quad F = 0 \end{aligned} \qquad 49$$

It appears that we have here once again a case of viscous flow with a variable viscosity dependent now on the current stresses. It is however possible to reduce the problem to that discussed in the previous section where viscosity is a function of the strain rate (as in eq. 41).

To do this we shall find it convenient to use a tensorial notation and introduce the concept of deviatoric stress. Thus eq. 9 defining viscosity becomes

$$\underline{S}_s = \underline{\sigma} - \begin{Bmatrix} 1 \\ 1 \\ 1 \\ 0 \\ 0 \\ 0 \end{Bmatrix} \sigma = \underline{D}\,\underline{\dot{\varepsilon}} \qquad 9$$

or in tensor notation

$$S_{ij} \equiv \sigma_{ij} - \delta_{ij}\,\sigma_{ij} = 2\mu\dot{\varepsilon}_{ij} \qquad 50$$

Taking the yield conditions defined simply by the second invariant of S_{ij} - i.e. the so called von Mises yield criterion

$$F = \sqrt{\tfrac{1}{2} S_{ij}\,S_{ij}} - Y \qquad 51$$

in which Y is the yield stress in simple tension, we find that

$$\frac{\partial F}{\partial \sigma_{ij}} = \frac{\partial F}{\partial S_{ij}} = \frac{1}{\sqrt{\tfrac{1}{2} S_{ij}\,S_{ij}}}\, S_{ij} \qquad 52$$

and that we can write eq. 48b as

$$\dot{\varepsilon}_{ij} = \frac{1}{2\bar{\mu}} < \sqrt{\tfrac{1}{2} S_{ij}\,S_{ij}} - Y >^{n} . \quad \frac{1}{\sqrt{\tfrac{1}{2} S_{ij}\,S_{ij}}}\, S_{ij} \qquad 53$$

Comparison with the definition of viscosity in eq. 50 gives

$$\frac{1}{\mu} = \frac{1}{\bar{\mu}}\, \frac{< \sqrt{\tfrac{1}{2} S_{ij}\,S_{ij}} - Y >^{n}}{\sqrt{\tfrac{1}{2} S_{ij}\,S_{ij}}} \qquad 54$$

but using 50

$$\sqrt{\tfrac{1}{2} S_{ij}\,S_{ij}} = \mu \sqrt{2\,\dot{\varepsilon}_{ij}\,\dot{\varepsilon}_{ij}} = \mu\,\bar{\dot{\varepsilon}} \qquad 55$$

Inserting this in above we see that μ can be obtained in terms of $\bar{\dot{\varepsilon}}$. For n=1 it can easily be verified that

$$\mu = \frac{Y + \bar{\mu}\,\dot{\bar{\varepsilon}}}{\dot{\bar{\varepsilon}}} = (\dot{\bar{\varepsilon}}) \qquad 56$$

and the general form identical with that discussed in previous section is obtained.

Computationally the expression 56 appears to present a difficulty with $\mu \to \infty$ as $\dot{\bar{\varepsilon}} \to 0$. However this problem is readily overcome by limiting the upper value of μ to some large number.

The viscoplastic model used is one of a class suggested by Perzyna (13,14) and is of quite wide applicability. It is interesting to note that as the coefficient $\mu \to 0$ the viscoplastic and plasticity formulation become identical. The solution procedure is therefore applicable to problems of both plastic and viscoplastic kind.

5.4 Inclusion of dynamic effects - Navier Stokes problem

It was shown in section 2 how the inclusion of the dynamic term in the virtual work discretisation gave rise to a non-linearity and to non-symmetric matrices. The lack of symmetry is due to the absence of a variational principle for full viscous flow problems.

The problem presented is known to have unique solution at low Reynolds numbers but for large Reynolds number steady state solution apparently do not exist and the flow becomes turbulent - a fact found experimentally. Solution to the steady state problem will thus be only sought for fairly low velocity flows. With the transient term $\frac{\partial u}{\partial t}$ (vide eq. 5) included in principle solution could be obtained for any velocity but due to the nature of turbulence and its rapid velocity changes numerical errors could well be expected unless both time and space subdivisions are very fine.

In the low velocity steady state problem we can proceed numerically in two ways: Either by isolating the non-linearity in the "force" term and writing equations corresponding to 24 as

$$\underline{\bar{K}}_o\, \underline{a} + \underline{\bar{f}} = 0 \qquad 57$$

in which $\underline{K}_o$ is the symmetric, constant, matrix and $\bar{f}$ is dependent on velocity and hence

$$\underline{\bar{f}} = \bar{f}(\underline{a}) \qquad 58$$

An iteration of the type

$$\underline{a}^{n} = \underline{K}_{o}^{-1} \underline{f} (\underline{a}_{n-1}) \qquad 59$$

is effective however only at very low velocities and at higher ones does not converge. Alternatively we use the non-linear form

$$\underline{K}\,\underline{a} + \underline{f}_{o} = 0 \qquad 60$$

in which

$$\underline{K} = \underline{K} (\underline{a}) \qquad 61$$

and iterate as

$$\underline{a}^{n} = \underline{K} (\underline{a}_{n-1}) \underline{f}_{o} \qquad 62$$

This process has proved effective for quite high velocities (Reynolds numbers) but is obviously more costly requiring repeated inversion of a non-symmetric matrix which has to be recalculated at each stage.

Formulations used so far in this problem include both the approaches outlined in sections 2.2 and 2.3 i.e. the use of velocities and pressures as unknown or the use of stream function (in two dimensional situations).

Reference 3 shows some problems solved using the first type of formulation and an interpolation of C_o continuity using isoparametric, parabolic elements.

It is found that in this type of formulation an improvement of results is achieved if a lower order of expansion is used for the variable $\underline{p}$ than for the velocities $\underline{u}$. Reasons for this are due to the order in which the variables occur in appropriate equation.

6 FLOW OF INVISCID FLUIDS

6.1 Equations of motion

The special case of flow of 'perfect' incompressible fluid has been left to the end of this chapter although it is by far the simplest problem to discretise and solve. (Indeed the formulation given by the author as early as 1965 provided the basis for many subsequent solutions (15)).

The reason for this is that it is reasonably simple to degenerate the more general equations given for the case of zero viscosity.

If we examine eq. 9/10 we note that the stress $\underline{\sigma}$ can not have shear components and is now simply represented by a pressure p i.e.

$$\underline{\sigma} \equiv - \begin{Bmatrix} 1 \\ 1 \\ 1 \\ 0 \\ 0 \\ 0 \end{Bmatrix} p \qquad 63$$

Equilibrium equation 3 becomes now

$$- \frac{\partial p}{\partial x} + X = 0$$

$$- \frac{\partial p}{\partial y} + Y = 0 \qquad 64$$

$$- \frac{\partial p}{\partial z} + Z = 0$$

inserting explicitly the body forces, i.e.

$$\underline{b} = \underline{b}_o - \rho \, \underline{a} = [X, Y, Z]^T \qquad 65$$

with the acceleration expression 5 and further assuming that the prescribed body forces b_o can be given by a potential Ω , i.e.

$$X_o = - \frac{\partial}{\partial X} \qquad \text{etc.} \qquad 66$$

or we can rewrite the equilibrium statement as

$$\frac{\partial p}{\partial X} + \frac{\partial \Omega}{\partial X} + \rho \left(\frac{\partial u}{\partial t} + u \frac{\partial u}{\partial x} + v \frac{\partial u}{\partial y} + w \frac{\partial u}{\partial z} \right) = 0 \qquad 67$$

etc....

Using vector notation the above set can be written as

$$\frac{1}{\rho} \nabla p + \frac{1}{\rho} \nabla \Omega + \frac{\partial}{\partial t} \underline{u} + (\underline{u} \cdot \nabla) \underline{u} = 0 \qquad 68$$

and observing a vector identity

$$\tfrac{1}{2} \nabla(\underline{u} . \underline{u}) = (\underline{u} . \nabla)\underline{u} + \underline{u} \text{ X curl } \underline{u} \qquad 69$$

an alternative form is conveniently available

$$\frac{1}{\rho} \nabla p + \frac{1}{\rho} \nabla \Omega + \frac{\partial \underline{u}}{\partial t} + \tfrac{1}{2} \nabla(\underline{u} . \underline{u}) - \underline{u} \text{ X curl } \underline{u} = 0 \qquad 70$$

In the last form of equilibrium relation a term known as vorticity occurs defined as

$$\underline{w} = \text{curl } \underline{u} \quad \left\{ w_x = \frac{\partial}{\partial y} u - \frac{\partial}{\partial x} v \quad \text{etc.} \right\} \qquad 71$$

It can be shown that in an inviscid fluid vorticity can not be created if flow is started from rest (see Batchelor, G. K., An Introduction to fluid dynamics, Cambridge U.P. 1967, p. 273, Kelvin's circulation theorem). Considerable interest centres therefore on so called irrotational flow in which

$$\underline{w} = 0 \qquad 72$$

For steady state motion of such a fluid equation 70 can be integrated yielding for ρ constant (incompressible fluid)

$$\frac{p}{\rho} + \frac{\Omega}{\rho} + \underline{u}.\underline{u} = \text{const.} \qquad 73$$

which is the well known Bernoulli equation.

6.2 Kinematics of irrotational and incompressible flow

To satisfy irrotationality condition 72 we simply need to define velocity by a potential ψ i.e.

$$u = -\frac{\partial \psi}{\partial x} \quad \text{etc.} \qquad 74$$

or $\underline{u} = -\nabla\psi$

with curl $\nabla\psi \equiv 0$

the vorticity is identically zero.

Further if the flow is incompressible then

$$\text{div } \underline{u} = \frac{\partial u}{\partial x} + \frac{\partial v}{\partial y} + \frac{\partial w}{\partial z} = 0 \qquad 75$$

and the potential has to obey the Laplacian equation

$$\nabla^2\phi = \frac{\partial^2\phi}{\partial x^2} + \frac{\partial^2\phi}{\partial y^2} + \frac{\partial^2\phi}{\partial z^2} = 0 \qquad 76$$

If the velocities (and hence the gradients of ψ) are known on the boundaries of the region the flow pattern is defined by the unique solution to eq. 76. Thus kinematics alone determine completely the velocity pattern.

If pressure distribution is needed the simple application of the Bernoulli's eq. 73 allows this to be found.

Alternatively we can proceed, in two dimensional problems by ensuring the incompressibility by defining a stream function (or for three dimensions a vector potential as shown in Sect. 2.3). Now the condition that $\underline{w}$ = 0 yields a governing equation

$$\nabla^2\psi = 0 \qquad 77$$

with appropriate boundary conditions for two dimensional problems (or

$$\text{curl. curl } \underline{\psi} = 0$$

in three dimensions when ψ is the vector potential).

Again kinematics defines the flow completely if velocities are prescribed.

We have discussed in detail the formulation of this simple equation 76 or 77 in the context of the finite element method and hence repetition is unnecessary here except to mention some special problems which may occur.

6.2 Special problems in numerical solution of irrotational flow

The first difficulty arises if free surface conditions exist. Such surfaces are streamlined and hence kinematic boundary conditions are known. However simultaneously the pressure is specified on the surface. This in fact means that the boundary position is undetermined a priori.

Solution generally proceeds in an iterative manner adjusting the boundary until both conditions are satisfied.

7 CONCLUDING REMARKS

In the lecture we have discussed a series of classical problems of fluid mechanics to show the application of the general finite element procedures. Many problems such as transient flow, compressible flow etc. have not been touched upon for lack of space although innumerable applications have already been made. For a complete bibliography the reader is referred elsewhere.

REFERENCES

1. O. C. ZIENKIEWICZ The Finite Element Method in Engineering Science, McGraw-Hill 1971.

2. L. R. HERRMANN Elasticity equations for incompressible or nearly incompressible materials by a variational theorem.

3. C. TAYLOR and P. HOOD. 'A Numerical Solution of the Navier-Stokes Equations using the Finite Element Technique' Computer and Fluids, Vol. I pp. 73-100. Pergamon Press 1973.

4. J. T. ODEN 'The finite element method in fluid mechanics' Lecture for NATO Advanced Institute on finite element method in Continuum Mechanics, Lisbon, Sept. 1971.

5. O. C. ZIENKIEWICZ and P. N. GODBOLE Incompressible elastic materials. A stream function approach to FEM solution. To be published In. J. N. Meth. Eng.

6. B. FRAEIJS DE VEUBEKE and O. C. ZIENKIEWICZ 'Strain energy bounds in finite elements by slab analogy' J. Strain Analysis 2 265-71, 1967.

7. G. SANDER 'Application of the dual analysis principle' Proc. IUTAM Symposium on High speed computing of elastic structures, Univ. of Liege 1970. p. 167-209.

8. B. TABARROCK A variational principle for the dynamic analysis of continua by hybrid finite element method. In. J. Solids Struct. 7 p. 251-268, 1971.

9. O. C. ZIENKIEWICZ Constrained variational principles and penalty function methods in finite element analysis. Conf. on Numerical Solution of Differential equations. Dundee 1972 (to be published Springer)

10a B. ATKINSON, M. P. BROCKLEBANK, C. C. M. CARD and J. M. SMITH Low Reynolds number developing flows A.I.Ch.E.J. 15 548-63 1969.

10b B. ATKINSON, C. C. M. CARD and B. M. IRONS Application of the finite element method to creeping flow problems. Trans. Inst. Ch. Eng. 48 p.276-84, 1970.

11. O. C. ZIENKIEWICZ and P. N. GODBOLE Flow of plastic and visco-plastic solids with special reference to extrusion and forming processes. To be published Int. J. N. Meth. Eng.

12. K. PALIT and R. T. FENNER
a) Finite element analysis of two dimensional slow non-Newtonian flows. A.I.Ch.E. Journal 18 p.1163-9, 1972.
also
b) Finite element analysis of slow non-Newtonian channel flow 18 p.628-33.

13. P. PERZYNA Fundamental problems in Visco-plasticity Recent advances in Appl. Mechanics. Academic Press N.Y. 9, pp. 243-377, 1966.

14. O. C. ZIENKIEWICZ and I. C. CORMEAU Visco-plasticity solution by finite element process. Archives Mechanics (Poland) 24 p. 873-89, 1972.

15. O. C. ZIENKIEWICZ and Y. K. CHEUNG Finite element solution of field problems. The Engineer p. 507-10, 1965.

APPLICATION OF FINITE ELEMENT METHODS TO STRESS ANALYSIS

Ivar Holand

Division of Structural Mechanics,
The Norwegian Institute of Technology,
Trondheim, Norway

ABSTRACT

The lecture starts with a historical review, drawing the attention to the pioneers Langefors, Argyris and Clough.

The partial differential equations encountered in problems of stress analysis are reviewed, and various sources of non-linearities are discussed. The relevant variational principles are mentioned and their application as a basis for finite element analysis is explained. Programming systems are discussed with major reference to the NORSAM system.

The merits of super-elements (or substructures) as related to input saving and equation solving are mentioned and illustrated by an example. Nonlinear problems (step by step loading, iteration) are treated briefly and illustrated by examples.

HISTORICAL REVIEW

To the best of our knowledge, the finite element method dates back to a paper by Courant in 1943 (1). It is interesting to note that the problems discussed by Courant concerned torsion, and thus belong to the field of stress analysis. Courant also announced a subsequent paper on plate bending by the same method, but this paper has never appeared.

Courant's approach was mathematically based. The term "finite element method" was not introduced until the middle of the fifties. At that time electronic computers were rapidly entering the field of technical computations, and matrix methods of structural analysis proved powerful. An extension of these methods to two- and three-dimensional solids was natural. Pioneers in this development were Langefors (2), Argyris (3) and Clough (4), and this time the approach was based on simple engineering arguments. The continuous material was regarded as being split physically into finite elements. Each element was analyzed as being a separate piece of material, making up the complete structure when joined to the other elements. The mathematical basis of the approach was not rediscovered and fully recognized until later.

PARTIAL DIFFERENTIAL EQUATIONS

A brief, but yet comprehensive description of the application of finite element methods in the theory of elasticity has been given by Pian and Tong in (5). For a thorough study, textbooks like the one by Zienkiewicz (6) should be recommended. Here, only a brief account of this theory may be included.

Elasticity problems are governed by three categories of field equations, viz.

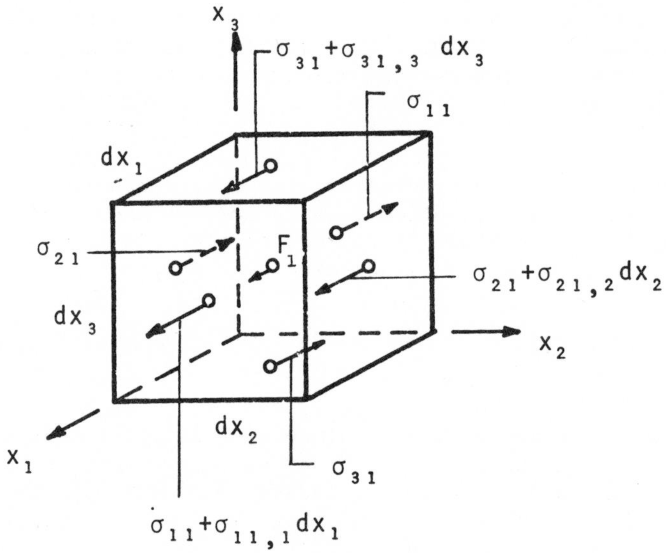

Fig 1. Three-dimensional state of stress, stresses in x_1-direction only are shown

stress equilibrium equations
stress-strain relations (constitutive material laws)
strain-displacement relations (kinematic relations)

In addition, boundary conditions may be given as

specified boundary stresses
specified boundary displacements

or, more generally, as specified relations between boundary stresses and boundary displacements.

In the case of linear theory of elasticity, these equations are particularly simple. In terms of rectangular Cartesian co-ordinates, and by use of standard tensor notation, they may be written

1. Stress equilibrium (Fig. 1)

$$\sigma_{ij,j} + F_i = 0 \qquad i = 1, 2, 3 \tag{1}$$

where the notations are

σ_{ij} stress tensor

F_i vector of volume forces

2. Stress-strain relations

$$\sigma_{ij} = C_{ijkl}E_{kl} \tag{2a}$$

or, inversely

$$E_{ij} = S_{ijkl}\sigma_{kl} \tag{2b}$$

where the new notations are

E_{ij} strain tensor

C_{ijkl} elastic stiffness coefficients

S_{ijkl} elastic flexibility coefficients

3. Strain-displacement relation for small displacements (Fig. 2)

$$E_{ij} = \tfrac{1}{2}(u_{i,j} + u_{j,i}) \tag{3}$$

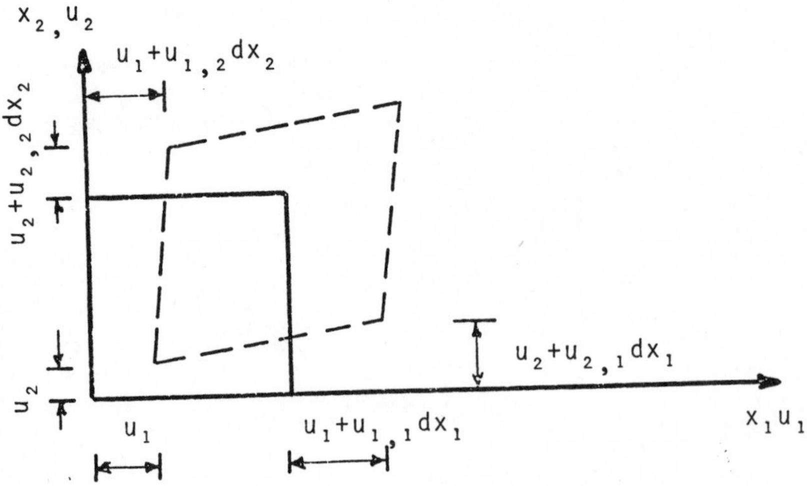

Fig 2. Strain-displacement relations in the plane x_1-x_2

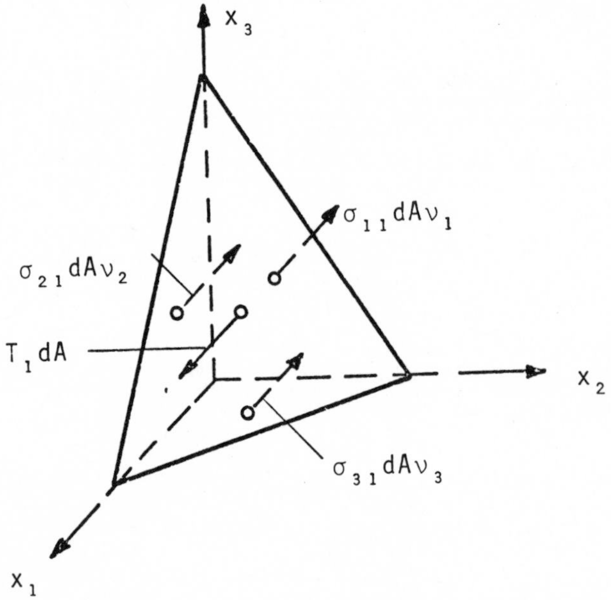

Fig 3. Surface tractions related to internal stresses. Stresses in x_1-direction only are shown

where u_i denotes displacement in direction i.

For the formulation of stress boundary conditions internal stresses must be related to surface tractions. The surface traction T_i in direction i at some part of the boundary S may be written (Fig. 3)

$$T_i = \sigma_{ij} \nu_j \tag{4}$$

where ν_j is the direction cosine of the surface normal.

This system of equations is linear, provided that the stiffness (or flexibility) coefficients are constant. One of the important advantages of the finite element method is to make analysis of nonlinear problems practicable, thus meeting needs of physical reality. The sources of non-linearities may be divided in two groups

material non-linearities
geometric non-linearities

The material non-linearities are described by coefficients C_{ijkl} or S_{ijkl} depending on the stress or strain, or on parameters like temperature, humidity, time etc.

The geometric non-linearities call for a careful consideration of the manner of describing stress and strain. The nature of the non-linearities appears clearly from the strain-displacement relation, the Green finite strain tensor being defined by

$$E_{ij} = \tfrac{1}{2}(u_{i,j} + u_{j,i} + u_{k,i}u_{k,j}) \tag{5}$$

The relation (3) is seen to represent a fair approximation only when the quadratic terms are negligible (small strain).

VARIATIONAL PRINCIPLES

Variational principles have been discussed already by Mitchell (7). The variational principles applicable in the case of small displacement theory shall be reviewed briefly in this context.

The principle of minimum potential energy states that the variation of the potential energy Π_p is zero, where

$$\Pi_p = \int_V (\tfrac{1}{2}C_{ijkl}E_{ij}E_{kl} - \overline{F}_i u_i)dV - \int_{S_\sigma} \overline{T}_i u_i dS \tag{6}$$

where V denotes the volume and S the surface of the body. In this case, the known quantities are the volume forces $\overline{F}_i$ and the surface tractions $\overline{T}_i$, which are assumed to be specified over a part S_σ of the boundary. The quantities to be adjusted so as to make the variation vanish are the displacement components u_i, by which also the strains E_{ij} are expressed. Thus, in an approximate solution strain-displacement (Eq. (3)) and stress-strain relations (Eq. (2a)) are satisfied identically, whereas stress equilibrium (Eq. (1)) is satisfied only approximately.

Because displacements are the primary unknowns a method based on the minimum potential energy principle is also called a displacement method.

In the principle of minimum complementary energy the functional to be minimized is

$$\Pi_c = \int_V \tfrac{1}{2} S_{ijkl}\sigma_{ij}\sigma_{kl}\,dV - \int_{S_u} T_i\overline{u}_i\,dS \tag{7a}$$

In this case, the known quantities appearing in the functional are the surface displacements $\overline{u}_i$, which are specified over a portion S_u of the surface. The quantities to be adjusted so as to make the variation vanish are the stresses σ_{ij}. The stress patterns chosen satisfy equilibrium identically, and strains may be computed according to Eq. (2b), but these strains do not in general satisfy the equations of compatibility, that is, they cannot be expressed by displacements in a unique manner, as given in Eq. (3). It is also possible to specify internal displacements if the prescription of volume loads is given up, see for instance (8). In this case the functional is generalized to

$$\Pi_c = \int_V (\tfrac{1}{2} S_{ijkl}\sigma_{ij}\sigma_{kl} - F_i\overline{u}_i)\,dV - \int_{S_u} T_i\overline{u}_i\,dS \tag{7b}$$

and the correspondence to Π_p (Eq. (6)) is more obvious. Because stresses are the primary unknowns, a method based on the complementary energy principle is also called a force (or equilibrium) method.

More general variational principles for stress analysis may be established by relaxing the requirements on displacement-strain compatibility in the minimum potential energy principle and the requirements on stress equilibrium in the minimum complementary energy principle. Washizu (9) presents the following generalized principle of minimum potential energy (except for notation)

$$\Pi = \int_V (\tfrac{1}{2}C_{ijkl}E_{ij}E_{kl} - \bar{F}_i u_i)dV - \int_V [E_{ij} - \tfrac{1}{2}(u_{i,j} + u_{j,i})]\, \sigma_{ij} dV -$$

$$\int_{S_\sigma} \bar{T}_i u_i dS - \int_{S_u} T_i (u_i - \bar{u}_i) dS \qquad (8)$$

The independent quantities to be varied in this functional are eighteen in number, viz.

6 strains	E_{ij}
3 displacements	u_i
6 stresses	σ_{ij}
3 surface tractions	T_i

Thus, neither strain-displacement compatibility nor stress equilibrium is satisfied in advance. In Eq. (8) σ_{ij} and T_i are Lagrange multipliers having the physical meaning specified by the notation used. Lagrange multipliers having been employed, the functional Π can only be required to be stationary, not a minimum as in Eq. (6).

The differential equations of stress analysis has so far been explained in terms of a threedimensional theory of elasticity. This is convenient as far as general theorems and description of the procedure are concerned. For engineering purposes, however, simplified structural behaviour must be considered, to make even modern computer analyses practicable. The structural elements in question are, arranged according to increasing complexity

straight beam
curved beam
flat plate, loaded in plane stress (or strain)
flat plate, loaded in bending
curved plate or shell

The analysis of such structural elements is based on assumptions of various degrees of complexity. The simplest one is the Navier-Bernoulli hypotheses of plane sections remaining plane, as used in the engineer's theory of beams and in the theory of bending of thin plates and shells. The differential equations, and the variational principles are in such cases formulated in terms of stress resultants, displacement of center lines and so on. But in principle the equations used are the same as those shown for the three-dimensional case.

FINITE ELEMENT APPROXIMATIONS

All anergy principles described may be used as a basis for numerical analysis by the finite element method. The finite element discretization implies a division of the total volume V into subvolumes or subdomains denoted finite elements. The functions chosen to represent approximate displacement and stress fields are specified within each element, and conditions imposed on certain parameters at interelement boundaries provide the necessary continuity of field functions. The approach has been explained already in the lecture by Collatz (18). Various element types have been discussed by Mitchell (19).

In the case of the standard displacement method displacement at chosen modes located at element bounderies represent the displacement field. All nodal displacement components may be arranged in a vector $\mathbf{r}$. Completion of the integration in Eq. (6) gives the potential energy functional in the form

$$\Pi_p = \tfrac{1}{2}\mathbf{r}^T\mathbf{K}\mathbf{r} - \mathbf{r}^T\mathbf{R} \tag{9}$$

where

$\mathbf{K}$ is the stiffness matrix resulting from the first (quadratic) part of the volume integral

$\mathbf{R}$ is the load vector resulting from the surface integral and the last part of the volume integral.

Equating the variation of Π_p to zero, yields the linear system of equations

$$\mathbf{K}\mathbf{r} = \mathbf{R} \tag{10}$$

One of the main features of the finite element method is that the repetition of elements throughout the volume makes it possible to establish the stiffness matrix $\mathbf{K}$ and the load vector $\mathbf{R}$ in a very systematic manner.

Fig. 4 shows the principle of adding the contribution from an element No. α into Eq. (10). A triangular element with 3 nodes is chosen for illustration. The assembly process consists in replacing local numbers by global ones and adding the submatrices of the element in the corresponding places of the global matrix.

Proofs of convergence of the method are closely related to interelement continuity of the assumed functions for the field variables. Lack of continuity may be compensated by use of mixed variational principles like (8), where discontinuities along

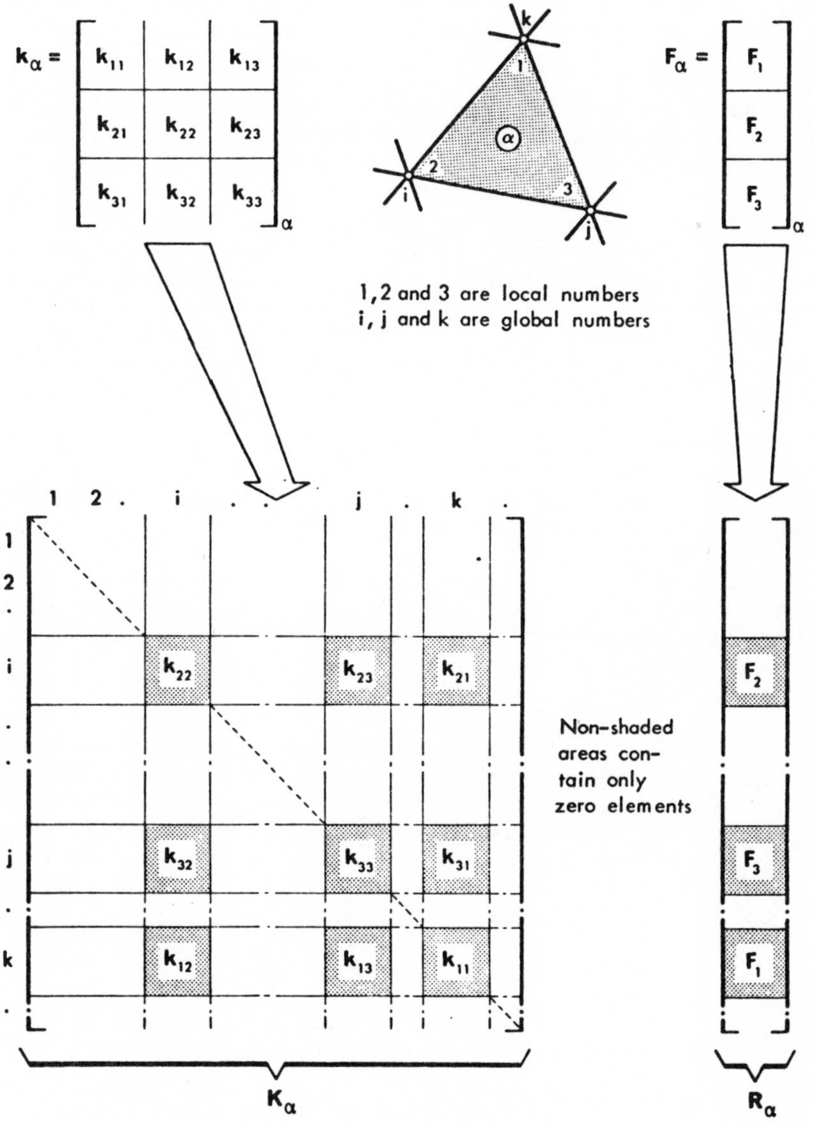

Fig 4. Stiffness and load merge

element boundaries may be minimized by including Lagrange multipliers.

PROGRAMMING SYSTEMS

Even though the finite element method is ideally suited for computer programming because of its systematic approach, the necessary statements are numerous and the error sources are manyfold. Hence, much effort has been put into developing special and general purpose programs, that could be used for the analysis of a large variety of physically very different-looking problems. Examples of general programs are ASKA (10), NASTRAN (11) and SESAM 69 (12). There are of course all possibilities of varying the size and complexity of such general programs. The biggest so far is probably NASTRAN, which comprises about 200.000 FORTRAN cards. NASTRAN will be discussed further by Spreeuw (20).

A system called NORSAM (13) has been developed by a Norwegian team representing industry and research institutions, with a substantial economic support from the Royal Norwegian Council for Scientific and Industrial Research.

NORSAM is intended to be, not a general purpose program, but a tool-kit for the advanced programmer of special purpose finite element analysis programs. It consists of approximately 250 FORTRAN routines, 18000 executable FORTRAN statements and 30000 punched cards.

The idea and the scope of the system is possibly best illustrated by the system chart shown in Fig. 5. The NORSAM system is divided into 10 subsystems (numbered 0 to 9) as indicated by the various boxes of Fig. 5. The modules of the various subsystems perform the following operations

0 - Data Handling System (DHS) and general service routines

Modules primarily concerned with:

- transfer of data between core (central memory) and peripheral storage;
- allocation of data storage on peripheral storage;
- saving of data on to permanent peripheral storage (restart facilities).

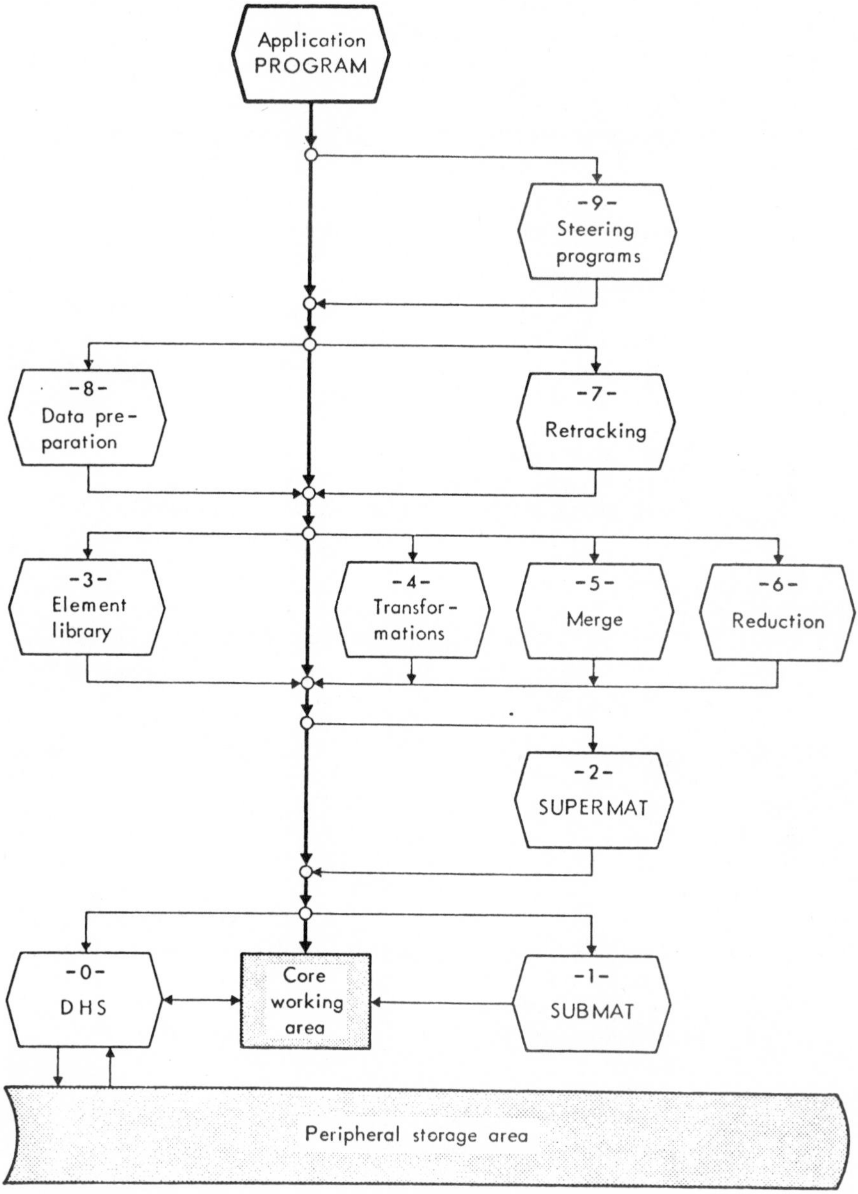

Fig 5. NORSAM system chart

1 - SUBMAT

Modules performing various matrix operations on submatrices (in-core matrices), ranging from generation of zero matrices to determination of eigenvalues and eigenvectors.

2 - SUPERMAT

Modules performing various matrix operations on supermatrices (out-of-core matrices).

3 - Element library

Modules for generating matrices (stiffness, stress, load, mass etc.) for various kinds of BASIC elements.

4 - Transformations

Modules for generation of TRANS elements, by transformation of existing elements.

5 - Merge

Modules for assembling (merging) subelements into substructures (formation of the systemmatrices **K**,**M**,**C**,**R**).

6 - Reduction

Modules for reduction (elimination) of internal degrees of freedom.

7 - Retracking

Modules for computation of displacements/reactive forces/stresses.

8 - Data preparation

Modules for reformatting of data (from user's format to NORSAM format), and standard modules for assisting an automatic data generation and graphic plotting.

9 - Steering programs

Higher level modules (not yet defined).

The numbering of the subsystems is such that a module of a particular subsystem may, in general, reference any module from its own subsystem or from subsystems with lower numbers. This is not a built-in limitation, it is merely a consequence of a logical grouping. All NORSAM modules do, however, comply with this rule. The subsystems may therefore, logically, be arranged on different levels as indicated schematically by the system chart shown in Fig. 5. Modules from one "box" (program or subsystem) may reference modules from other boxes as indicated by the arrows.

In problems of stress analysis the system of equations is symmetric, banded and positive definite. In modern finite element analysis the direct methods of solution are dominating. Nevertheless, iterative methods as discussed by Young (14) may be used to advantage in particular cases, e.g. in combination with a direct method in nonlinear problems. Hence, both methods should be included in a general programming system. However, direct methods should have the higher priority, and the storage format, which in NORSAM is a submatrix scheme, was chosen with the direct method in mind. A standard Cholesky triangular factorization algorithm based on hypermatrices is used. It takes, however, advantage of sparseness in a slightly unusual manner. In principle, it works with the entire upper triangular part of the original matrix, but only non-zero submatrices are operated on. Thus the effectiveness of the algorithm is fairly independent of the node numbering (bandedness or not bandedness).

The iterative solution in NORSAM is based on the method of conjugate gradients.

A very important feature of NORSAM is the independency of the special problem considered. Thus, its use is not limited to problems of stress analysis. It can equally well be adapted to the solution of problems of temperature distribution fluid flow etc. A limitation, but not a severe one, is that the system of equations must be symmetric and positive definite. The system can be modified for non-positive-definiteness.

SUBSTRUCTURES AND SUPERELEMENTS

Finite element analysis of complex structures very often lead to problems of such a size (several thousand degrees of freedom) that it is desirable to break down the structure in a set of substructures. This is particularly so if there is some sort of repetition of geometry in the structure. The parts in which the structure is broken down are denoted substructures. The advantages obtained by substructuring are of two categories. Firstly, the amount of input may be reduced, and datachecking made easier. In the case of several similar substructures, topological and elastic data may be given only once for each type of substructure. Secondly, computing may be reduced by eliminating only once all internal degrees of freedom in each substructure, leaving only those degrees of freedom which connect the substructure to the rest of the structure to be determined by the final system of equations. Computing time may be reduced due to the reduction of bandwidth, even in cases where the total number of operations should not be reduced due to substructuring. Similarly, starting from the element level, it is of advantage to assemble basic elements to groups of elements, called superelements. The difference between substructure and superelement is thus immaterial. It is only a matter of point of view in the specific case if some group of elements is denoted "substructure" or "superelement".

The procedure may be illustrated by an example described in (16). The problem concerns analysis of propagation of cracks. The energy release rate as the crack opens shall be computed, as this rate in turn can be used for estimating stress intensity factors. The strain energy U is found as

$$U = \frac{1}{2}\mathbf{R}^T\mathbf{r} \tag{11}$$

Where $\mathbf{R}$ is the vector of externally applied loads, and $\mathbf{r}$ is the vector of displacements of the loaded points. These displacements are computed by the finite element method. The procedure which is rather favourable numerically, since the accuracy of computed energy is higher than for instance of stresses or individual displacements.

The specific example concerns a semielliptical crack in a thick plate as shown in Fig. 6. Because of double symmetry, only one quarter of the plate needs to be analyzed. This quarter is shown in the central lower part of Fig. 7, labelled "Final model level 3". The basic element is a 20-node isoparametric hexahedron shown at the top of the figure. The basic element is used to build the 4 superelements at level 1.

In the superelements all internal displacements are eliminated, thus leaving retained displacements on the surface only. Note that a fine mesh is used at the bottom of superelement level 1, type 1, which is intended to be located at the crack. At larger distances from the crack, however, a coarser mesh has been used. A major problem is the transition from the fine to the coarse mesh. This can be accomplished by distorting the hexahedrons drastically, for instance into triangular prisms as shown several places in the superelements at level 1, or by introducing linear constraints between nodal displacements. An alternative version of superelement level 1, type 1, is shown in Fig. 8.

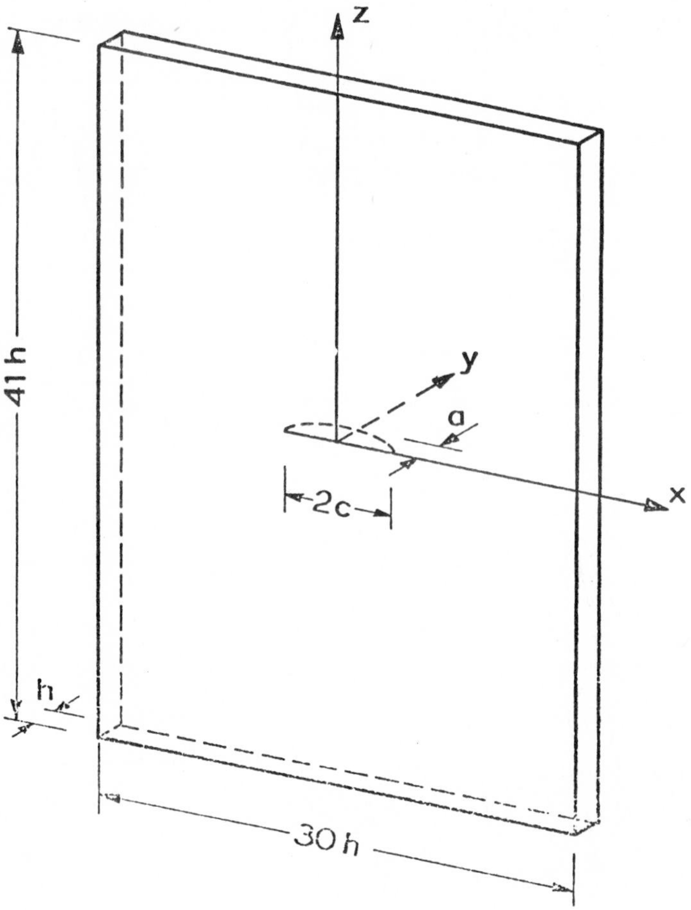

Fig 6. Semi-elliptical, part-through surface crack in a plate

Superelements at level 1 are combined to superelements at level 2 (lower left and right side of Fig. 7). The repetition of superelements in this case is limited, only superelement level 1, type 1 is used twice. Nevertheless, the superelement technique was felt to be highly beneficial, both in making data checking practicable, and in reducing actual computer time. Eventually, the level 2 superelements are combined to the final model, level 3.

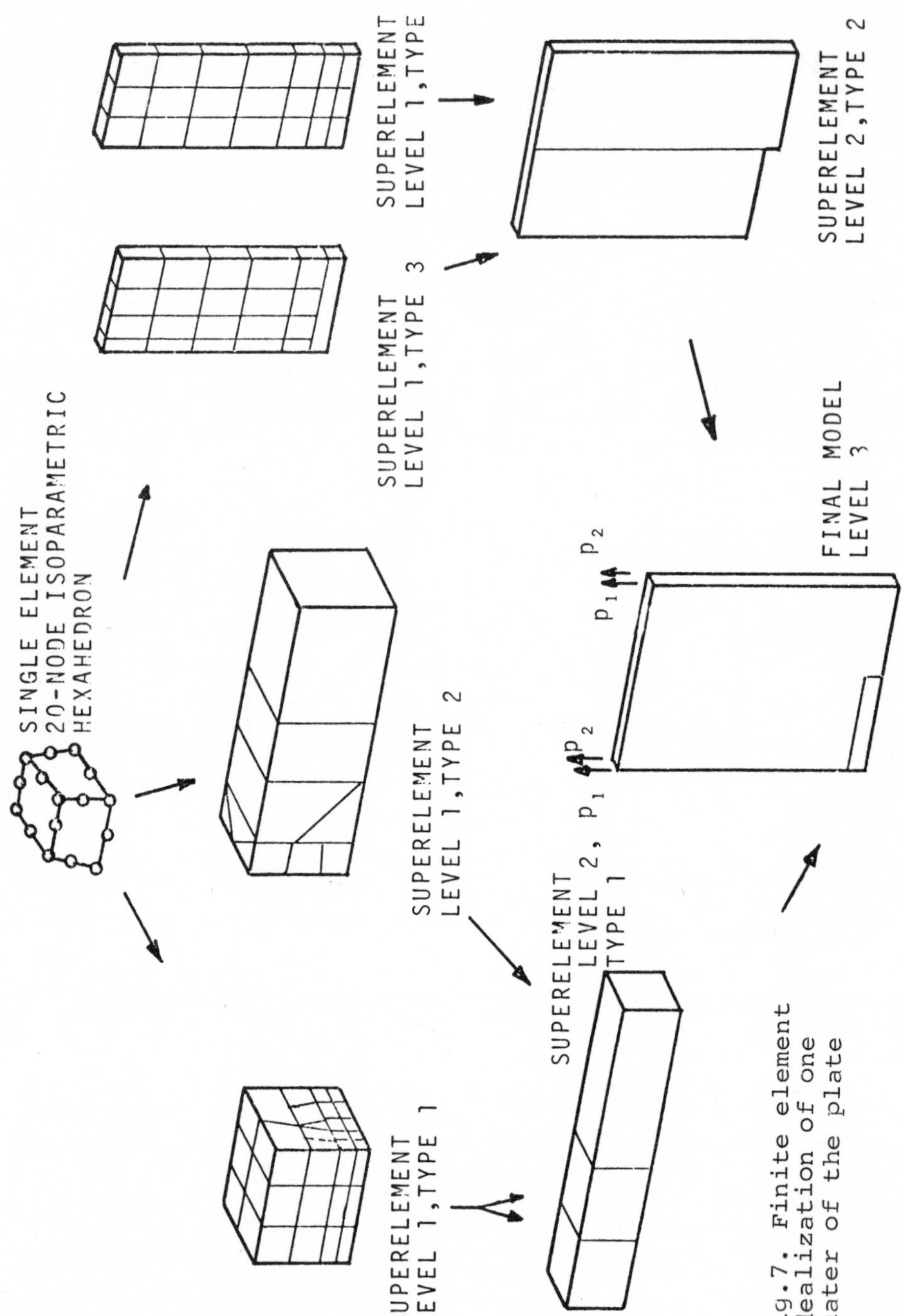

Fig.7. Finite element idealization of one quater of the plate

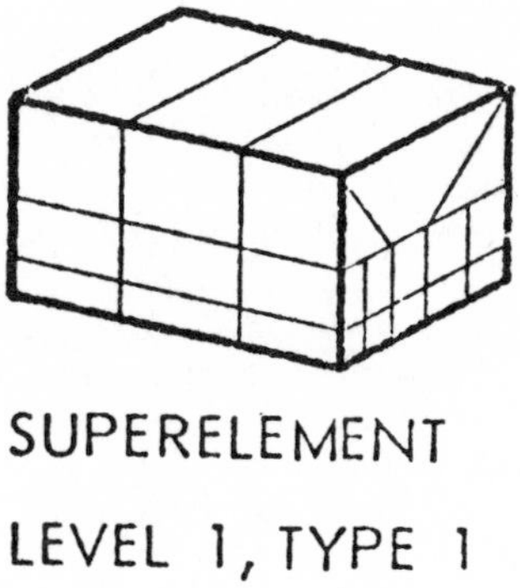

Fig 8. Alternative element idealization of the near crack superelement

The total model comprises 163 elements and 3360 nodal degrees of freedom. By successive elimination through the three levels only 4 nodal points were retained for load application, and 161 points, each with one degree of freedom, were retained in the plane of the crack. A total of approximately 2 hours of computer time was required for this reduction on a UNIVAC 1108 computer.

The remaining nodes in the crack plane of the final model are shown in Fig. 9. A great variety of different crack shapes were simulated by introduction of proper boundary constraints at

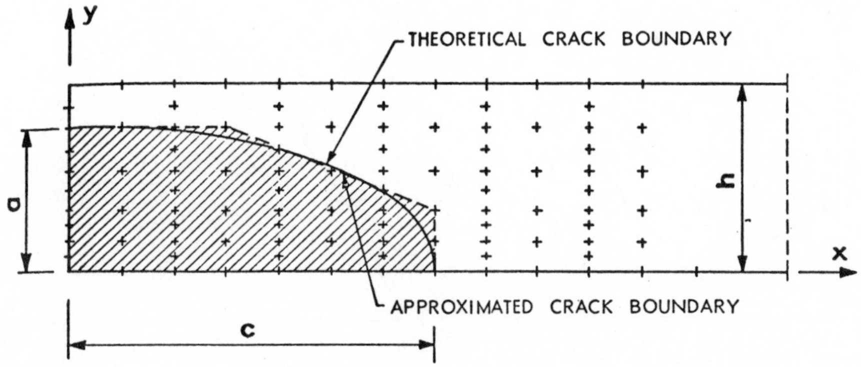

Fig 9. Remaining nodes in the x-y plane of the final model

the nodes in the plane of the crack. For instance, the elliptical crack in Fig. 6 was approximated by restraining the nodes on and outside the dashed boundary line from moving perpendicular to the plane, whereas the nodes inside this line were free to move. Since only very few degrees of freedom were retained in the final model, the analysis of every new crack shape for three loading cases required only 1.6 s CPU-time. Thus, by using this super-element technique analysis of a large number of crack geometries requires only insignificantly more computational effort than treating a single crack.

NON-LINEAR PROBLEMS

The finite element method has shown its power in modern numerical analysis because of its systematic ordering of numerical computations and because it lends itself readily to the solution of very complex problems.

This adaptability also permits the solution of nonlinear problems. One widely used procedure consists of loading the structure incrementally, linearizing the structural behaviour for every step (Fig. 10a). As this method alone may bring the computed path too far away from the correct solution, it should be supplemented by equilibrium corrections (Fig. 10b). The method is described in (6), and solution techniques for non-linear equations have been discussed by Professor Rall (17). Here, only a few examples of the application to stress problems shall be given.

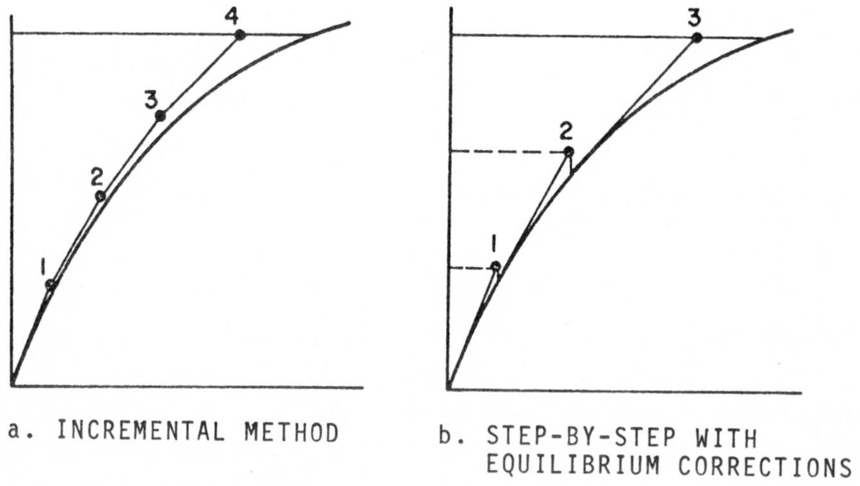

Fig 10. Non-linear analysis

The first example is not typical for non-linear analysis. It is included here to show an actual technical application. Furthermore, it serves to demonstrate that a rather crude representation of certain physical properties may suffice, provided that the results are not sensitive to the properties in question. The example concerns an analysis which was made for a new railway tunnel which is now being built through the central parts of Oslo. To avoid permanent reduction of the subsoil water pressure, the tunnel was provided with a concrete lining. As well concrete lining as rock behaves non-linearly. The actual material properties of the rock are difficult to ascertain, and they vary along the tunnel. Hence, it was felt most profitable to base the analysis on simple assumptions, and to investigate the influence of a wide variation of parameters. The tunnel section and a simple frame element model of the lining is shown in Fig. 11. In this model the rock was assumed to be perfectly rigid and the concrete perfectly elastic. The jointing surface between rock and lining was assumed to transfer compressive stresses only (no tensile or shear stresses). Thus the joint will have open portions and closed portions, the distribution of which depends on the deformations and is not known in advance. Thus, the problem is non-linear and must be solved by iteration. A plane strain model with a triangular element idealization is shown in Fig 12. The elements were of the linear strain type. In this model both concrete and rock were assumed to behave perfectly elastically, the source of non-linearity again being the joint, where the assumptions were: no transmission of stresses if the joint opens and full stress transmission if the joint closes. A completely rigid boundary was assumed in the rock 16 m away from the inner surface of the tunnel. Two extreme values of Young's modulus for the rock were considered, viz. E_r = 1000 MPa and E_r = 50 000 MPa. Young's modulus of the condrete was in both cases assumed to be E_r = 12 000 MPa. In all analyses only a few iterations were needed.

Fig. 13 shows the displacements computed for the boundary of the rock tunnel and for the outside of the lining. The figure demonstrates where contact between rock and linings occurs. The problem is of a type occurring in several physical applications, the characteristics being that the boundary between subdomains depends on the field variables, thus not being known in advance.

Results in terms of bending moments in the lining are shown in Fig. 14 for the three cases. The figure demonstrates that the results are similar, although the maximal values of bending moments differ significantly. In such non-sensitive cases, a simplified non-linear analysis as the one described may be sufficient.

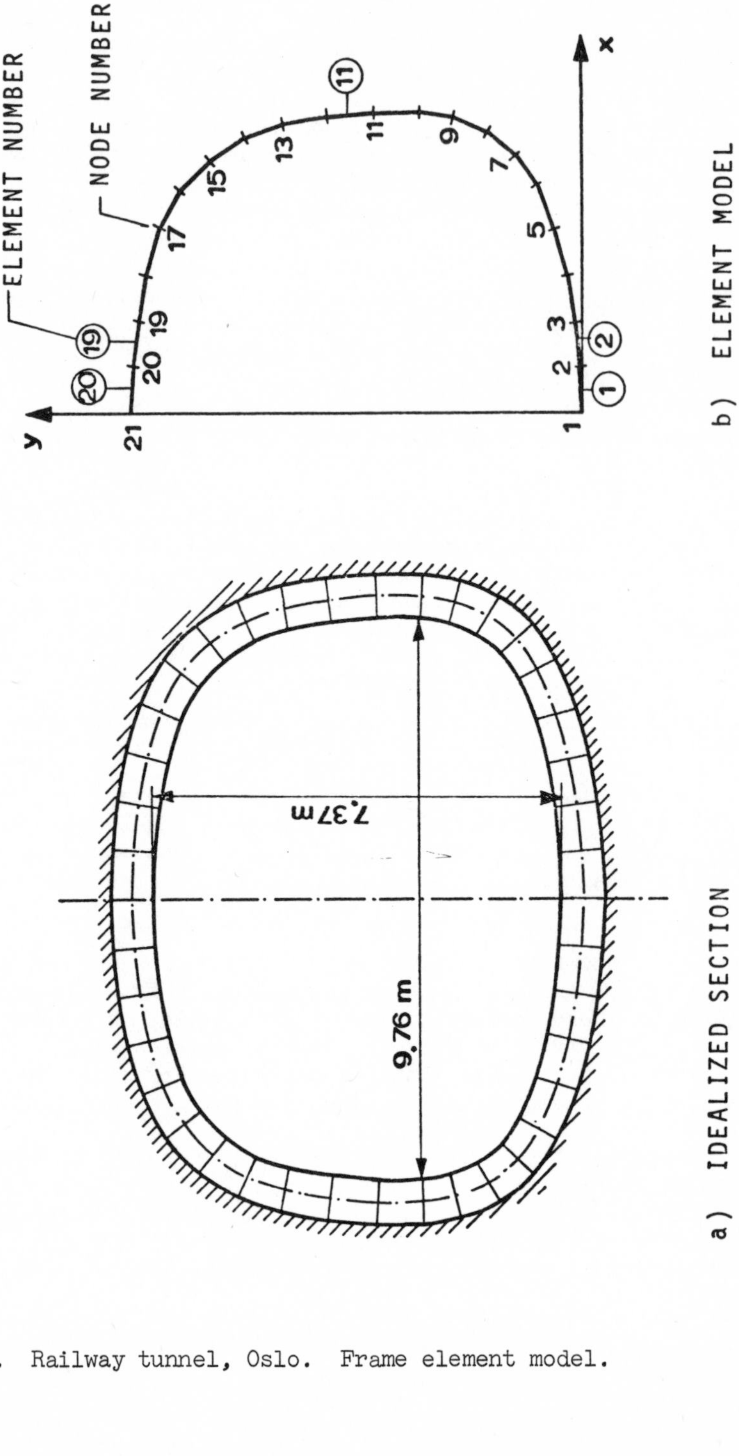

Fig 11. Railway tunnel, Oslo. Frame element model.

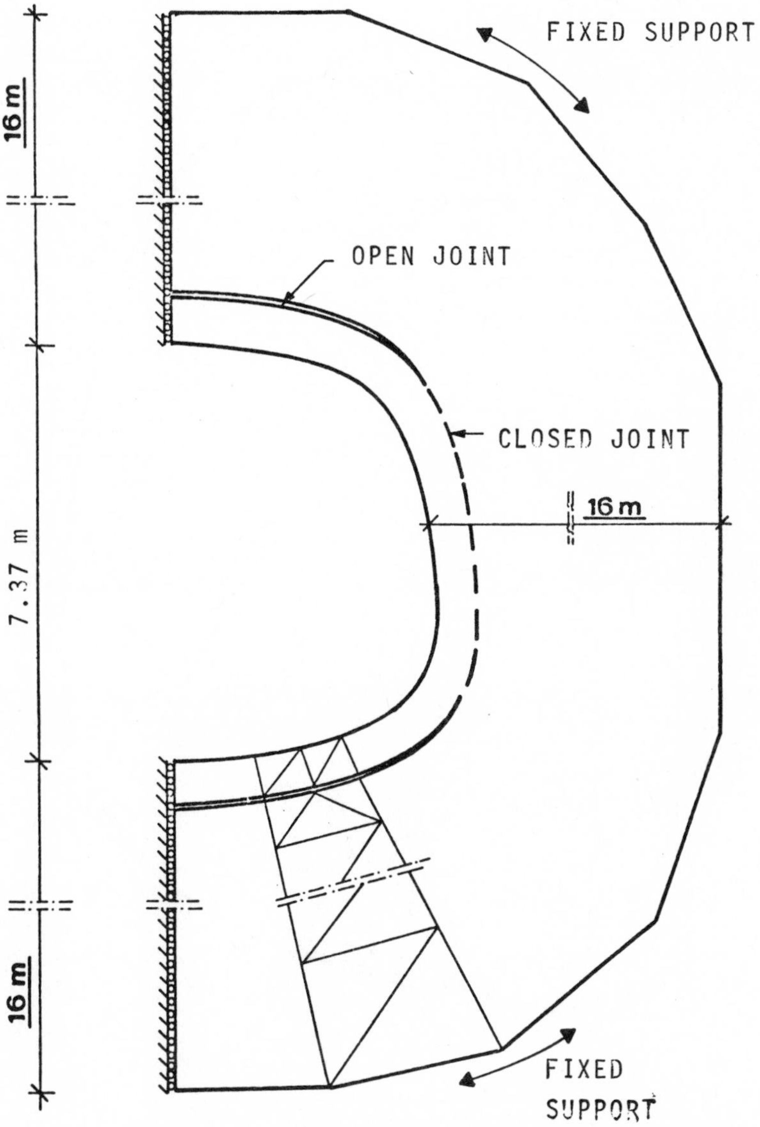

Fig 12. Railway tunnel Oslo. Plane strain element model.

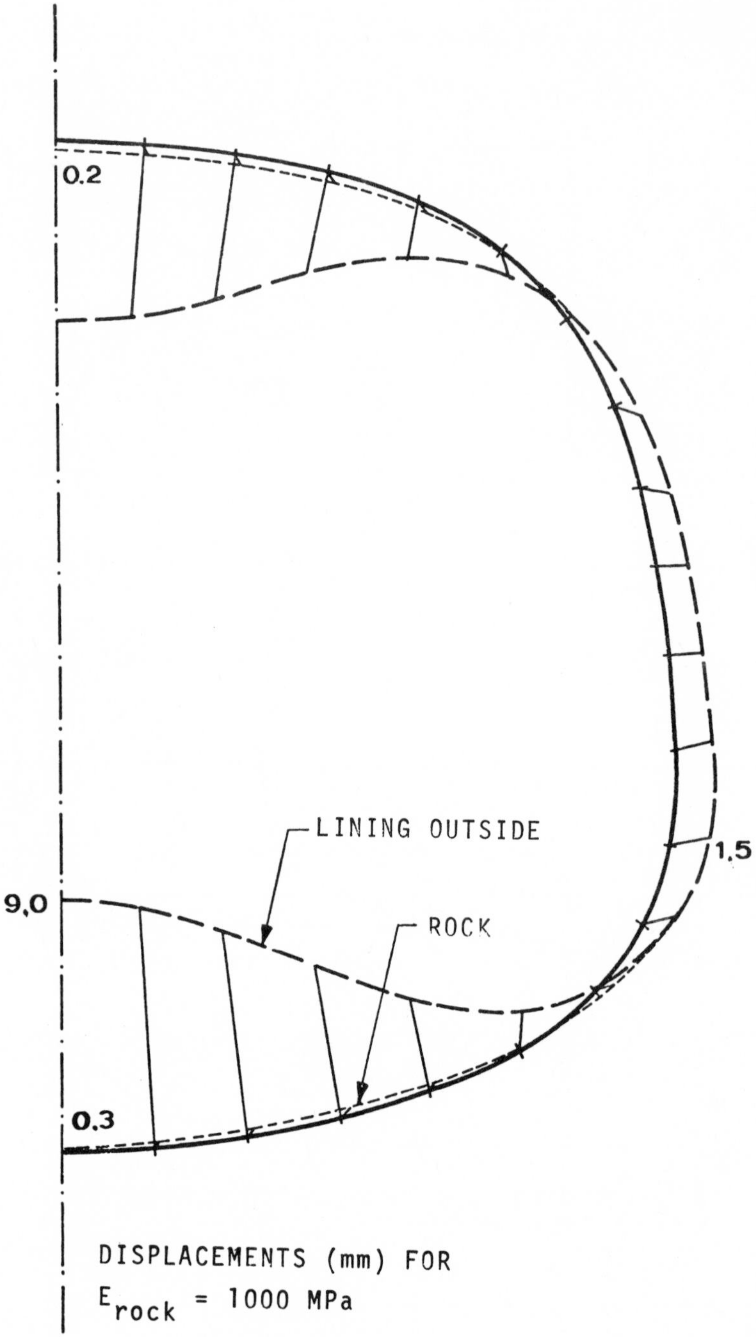

Fig 13. Railway tunnel, Oslo. Displacements.

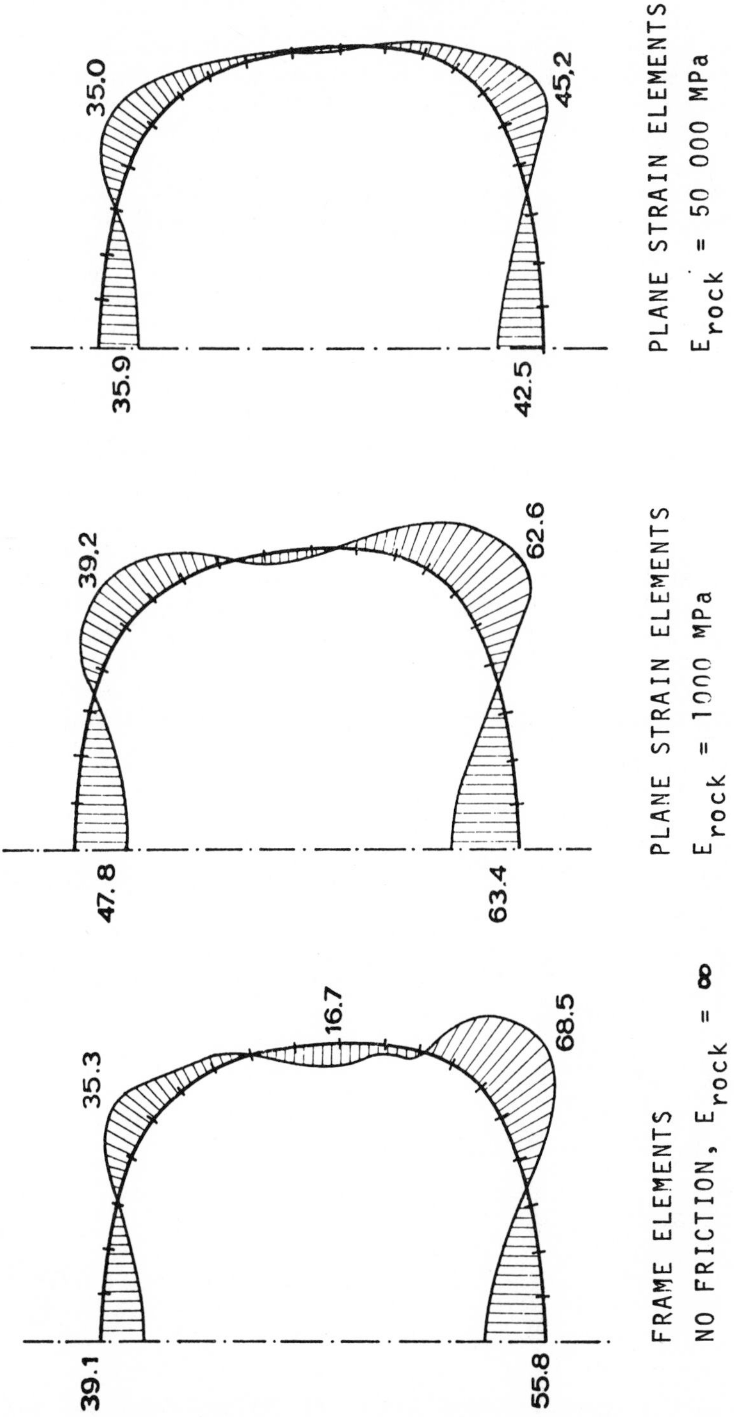

Fig 14. Railway tunnel, Oslo. Moments in concrete lining.

In other cases it is necessary to follow the actual behaviour of the structure as closely as possible during successive loading steps. As an example of such a procedure a study of non-linear behaviour of reinforced concrete plates (18) may be mentioned.

Reinforced concrete is composed of two basically different materials. The concrete behaves highly non-linearly in compression and is in many regions stressed in tension above the tensile strength, whereby cracking occurs. The reinforcement behaves approximately linearly up to a more or less well-defined yield stress. The transfer of stresses between steel and concrete, however, is non-linear, except at very low stress levels. An analytical model of this composite material was based on separate stress-strain diagrams for the two materials. The discrete reinforcement bars were modelled as a continuous membrane, with stiffness in the longitudinal direction of the bars only. The stress-strain relations of the concrete was described numerically on the basis of two-dimensional tests. The ultimate load for concrete plates is highly influenced by geometrically non-linear behaviour, which was incorporated according to the von Kármán theory. A plate, one quarter of which is shown in Fig. 15, had been tested for uniform lateral pressure. The plate was analyzed by using a two by two mesh of rectangular elements. 7 integration points were used through the thickness of the plate. An incremental stiffness was recomputed in each cycle of the equilibrium iteration. Fig. 16 shows the midpoint deflection compared with

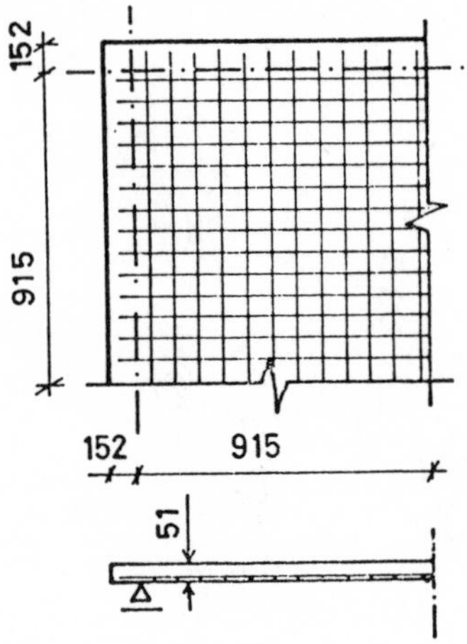

$f_c = 35\ N/mm^2$

Reinforcement 4.76 mm diameter plain round mild steel.
Yield stress 375 N/mm^2,
Ultimate stress 485 N/mm^2
Cover to the bottom layer 4.76mm

Fig 15. Simply supported slab tested by Taylor, Maher and Hayes

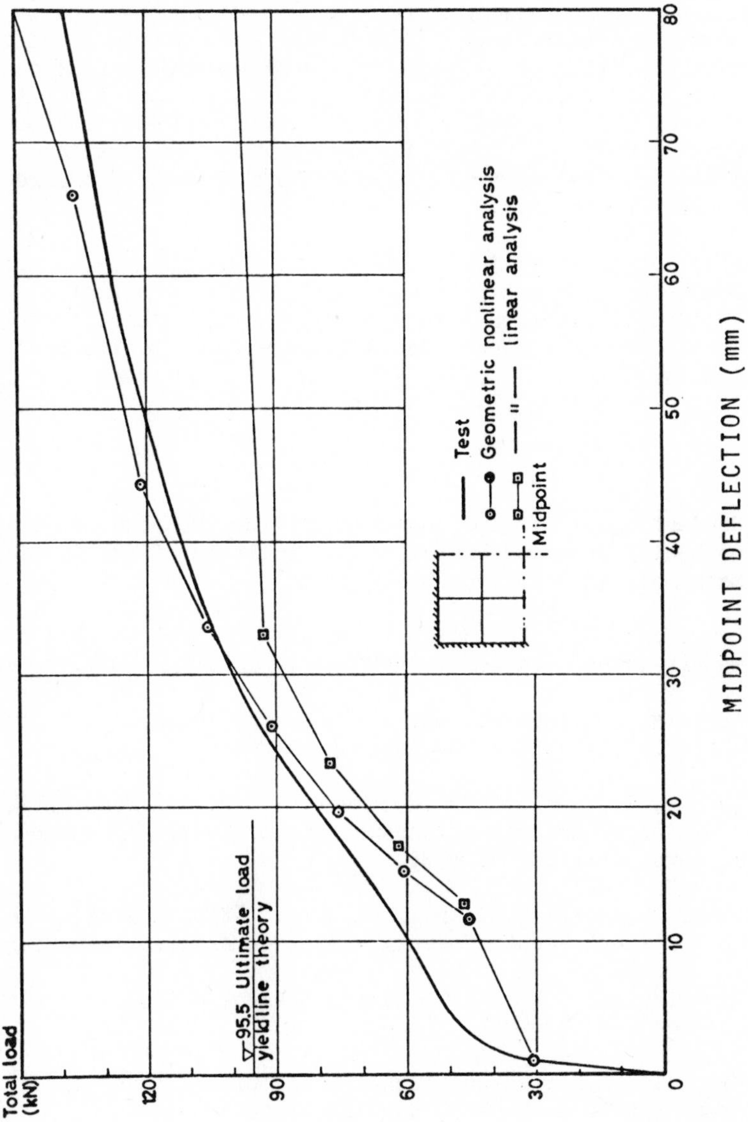

Fig 16. Load-deflection curves for simply supported square plate uniformly loaded

the test results. It is note-worthy that the discrepancies between experimental and analytical results are most pronounced immediately above the linear range. This is due to a not fully adequate description of the concrete behaviour immediately after cracking. A better description in a subsequent analysis gave a substantial improvement. In the region of high loads the geometric non-linearities prevail. These are more easily described accurately, resulting in a fair agreement. For further details the report (18) should be consulted.

Fig. 16 shows that the behaviour is highly non-linear, and should serve to demonstrate the capability of the finite element method to deal with non-linear problems. It is hard to believe that any other method can cope with non-linear problems of such a complexity.

REFERENCES

1. Courant, R.: "Variational Methods for the Solution of Problems of Equilibrium and Vibrations". Bulletin of the American Mathematical Society, 49, 1943, pp. 1-23.

2. Langefors, B.: "Structural Analysis of Swept-back Wings by Matrix-transformation", SAAB TN 3, Linköping 1951.

3. Argyris, J.H. and Kelsey, S.: "Energy Theorems and Structural Analysis", Aircraft Engineering, 1954-1955.

4. Turner, M.J., Clough, R.W., Martin, H.C. and Topp, L.C.: "Stiffness and Deflection Analysis of Complex Structures", J. Aero. Sci. 23, 9, 1956.

5. Pian, Theodore, H.H. and Tong, Pin: "Finite Element Methods in Continuum Mechanics". Advances in Applied Mechanics.

6. Zienkiewicz, O.C.: "The Finite Element Method in Engineering Science", McGraw-Hill, London 1971.

7. Mitchell, A.R.: "Variational Principles. A Survey". Present NATO Advanced Study Institute.

8. Fraeijs de Veubeke, B.: "Displacement and Equilibrium Models in the Finite Element Method", Stress Analysis, edited by O.C. Zienkiewicz and G.S. Holister, John Wiley & Sons Ltd., London 1965.

9. Washizu, K.: "Variational Methods in Elasticity and Plasticity", Pergamon Press, Oxford 1968

10. Schrem, E. and Roy, John R.: "An Automatic System for Kinematic Analysis, ASKA, Part 1", IUTAM Colloquium, Université de Liege, 1970.

11. MacNeal, R.H.: "The NASTRAN Theoretical Manual", NASA SP-221, Washington D.C. 1970.

12. Araldsen, P.O.: "SESAM-69. A General Purpose Finite Element Method Program", Application of Computerized Methods in Analysis and Design of Ship Structures, Marine Structures, and Machinery. Oslo 1972.

13. Bell, K., Hatlestad, B., Hansteen, O.E. and Araldsen, P.O.: "NORSAM. A Programming System for the Finite Element Method", User's Manual. February 1973.

14. Young, D.M.: "Solution of Linear Systems of Equations", Present NATO Advanced Study Institute.

15. Zienkiewicz, O.C.: "Application to Field Problems", Present NATO Advanced Study Institute.

16. Bergan, P.G. and Aamodt, B.: "Finite Element Analysis of Crack Propagation in Three-Dimensional Solids under Cycling Loading". To be presented at the 2nd International Conference on Structural Mechanics in Reactor Technology, Berlin 10-14th September 1973.

17. Rall, L.B.: "Solution of Nonlinear Systems of Equations", Present NATO Advanced Study Institute.

18. Berg, S.: "Nonlinear Finite Element Analysis of Reinforced Concrete Plates", Division of Structural Mechanics, The Norwegian Institute of Technology, Report No. 73-1, February 1973.

19. Collatz, L.: "Methods for Solution of Partial Differential Equations", Present NATO Advanced Study Institute.

20. Mitchell, A.R.: "Element Types and Base Functions", Present NATO Advanced Study Institute.

21. Spreeuw, E.: "Finite Element Computer Programs for Heat Conduction Problems", Present NATO Advanced Study Institute.

SOME FINITE DIFFERENCE METHODS FOR SOLUTION OF HEAT CONDUCTION PROBLEMS

E.E. Madsen and G.E. Fladmark

Institutt for Atomenergi, Kjeller, Norway

1. INTRODUCTION

The intention of this paper is to present some numerical methods for solution of the two-dimensional time-dependent heat conduction equation and to perform a numerical comparison between them both on linear and nonlinear problems.

The nonlinear effects included are temperature dependent coefficients and radiation. The coefficients are also considered piecewise constant in space.

In chapter 2 we present the mathematical model together with general boundary conditions. Chapter 3 deals with the space discretization and in chapter 4 we describe the four time-integration methods to be used. A numerical comparison concerning the stability, the error and the execution time on a CDC Cyber 74 is given in chapter 5 together with the conclusions.

2. MATHEMATICAL MODEL

Let τ be the domain of definition in the xy-space and $\partial\tau$ the boundary. Then the temperature distribution is given by:

$$\rho(\vec{r},T) \cdot c(\vec{r},T) \frac{\partial T(\vec{r},t)}{\partial t} = \nabla\lambda(\vec{r},T)\nabla T(\vec{r},t) + s(\vec{r},t,T) \tag{2.1}$$

$\vec{r}\in\tau$, $t \geq 0$

with the boundary condition

$$\alpha(\vec{r},T) \cdot T(\vec{r},t) + \beta(\vec{r},T) \frac{\partial T(\vec{r},t)}{\partial n} = \gamma(\vec{r},T) \tag{2.2}$$

$\vec{r} \in \partial\tau,\ t \geq 0$

and the initial condition

$$T(\vec{r},o) = To(\vec{r});\ \vec{r} \in \tau \tag{2.3}$$

where

ρ = density (kg/m^3)
c = specific heat (J/kg^oK)
λ = thermal conductivity (W/m^oK)
T = temperature (^oK)
t = time parameter (s)
$\vec{r}$ = position vector in space (m)
∇ = Laplacian operator
$\partial/\partial n$ = normal derivation operator
s = internal heat source (W/m^3)

The coefficients α, β and γ depend on the actual boundary conditions. However, they are assumed piecewise constant in space, and α and β should be non-negative. The following boundary conditions are actual in practice.

(i) Given temperature

$\alpha = 1,\ \beta = 0,\ \gamma$ = boundary temperature

(ii) Given heat flux

$\alpha = 0,\ \beta = \lambda,\ \gamma$ = boundary heat flux

(iii) Conduction

$\alpha = a,\ \beta = \lambda,\ \gamma = \alpha \cdot \theta$
where a denotes the thermal heat transfer coefficient for conduction and θ the temperature outside the domain ($\theta = \theta(\vec{r})$).

(iv) Convection

$\alpha = h,\ \beta = \lambda,\ \gamma = \alpha \cdot \theta$
where h denotes the thermal heat transfer coefficient for convection.

(v) Radiation

$\alpha = \epsilon \cdot \sigma \cdot (T + \theta)(T^2 + \theta^2)$, $\beta = \lambda$, $\gamma = \alpha \cdot \theta$
where ϵ denotes the emitivity factor and σ the Stefan Boltzmann constant.

A combination of the latter three types of conditions may occure at the same point. This is taken into account simply by adding the terms.

The coefficients ρ, c and λ are material dependent and therefore assumed piecewise constant in space.

The internal source may for example be caused by electrical resistance. Then it depends on both the temperature and the electric potential. However, in this paper we will ignore any internal heat source, because it is not essential in our discussions.

3. SPACE DISCRETIZATION

Let the domain τ be divided into rectangular or triangular cells with the mesh points represented by the corners, according to Fig 1.

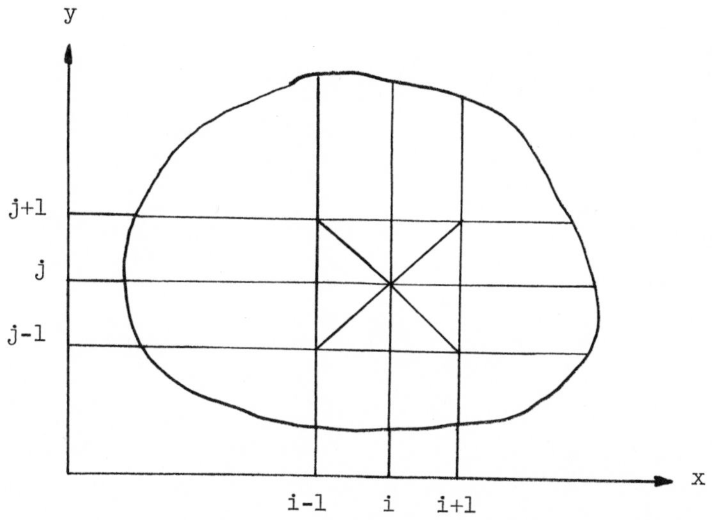

Fig 1. Division into cells.

The division should be done in such a way that each cell is homogeneous, which means that within a cell the material parameters ρ, c and λ are constant.

Using the notation given in Fig 2

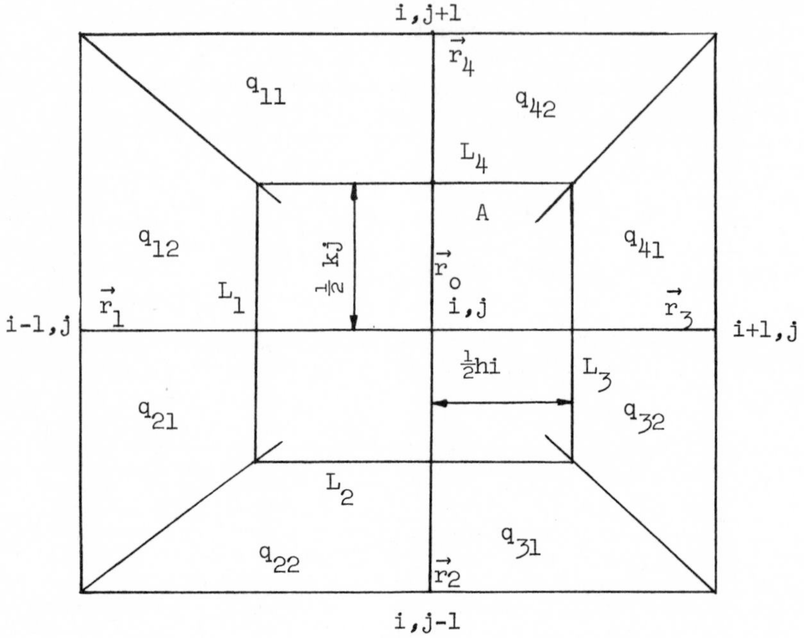

Fig 2. General situation of an internal mesh point

and assuming the mesh point (i,j) not to be a boundary point we have according to the Stokes theorem, ref (5):

$$\int_A \delta \frac{\partial T}{\partial t} \, dA = \int_A \nabla\lambda \nabla T dA =$$

$$= \int_{\partial A} \lambda \frac{\partial T}{\partial n} \, dL = \sum_{l=1}^{4} \int_{L_l} \lambda \frac{\partial T}{\partial n} \, dL \qquad (3.1)$$

with $\delta = \rho \cdot c$

The following approximations are now adopted:

1. $T(\vec{r},t) = T_{ij}(t) \; ; \; \vec{r} \in A$ (3.2)

2. $\frac{\partial T(\vec{r},t)}{\partial n} = \frac{1}{|\vec{r}_n - \vec{r}_o|} [T(\vec{r}_n,t) - T(\vec{r}_o,t)]; \; \vec{r} \in A$

where $\vec{r}_n$ is the nearest mesh point to $\vec{r}_o$ in $\vec{n}$ direction.

Then from eq (3.1) we have the following equation for mesh point (i,j):

$$\delta_{ij} \frac{d}{dt} T_{ij} = - \alpha_{oi,j} T_{ij} + \alpha_{1i,j} T_{i-1,j} + \alpha_{2i,j} T_{i,j-1} + \alpha_{3i,j} T_{i+1,j} + \alpha_{4i,j} T_{i,j+1} + S_{i,j} \quad (3.3)$$

where

$$\alpha_{1i,j} = \frac{1}{2\, h_{i-1}} (\lambda_{q12} k_j + \lambda_{q21} k_{j-1})$$

$$\alpha_{2i,j} = \frac{2}{2\, k_{j-1}} (\lambda_{q22} h_{i-i} + \lambda_{q31} h_i)$$

$$\alpha_{3i,j} = \alpha_{1\; i+1,j} \quad (3.4)$$

$$\alpha_{4i,j} = \alpha_{2\; i,j+1}$$

$$\alpha_{oi,j} = \sum_{\ell=1}^{4} \alpha_{\ell\; i,j}$$

$$\delta_{ij} = \frac{1}{8} [k_j h_{i-i} (\delta_{q11} + \delta_{q12}) + k_{j-1} h_{i-1} (\delta_{q21} + \delta_{q22}) + k_{j-1} h_i (\delta_{q31} + \delta_{q32}) + k_j h_i (\delta_{q41} + \delta_{q42})] \quad (3.5)$$

$$S_{i,j} = 0$$

The equation for a boundary point may be constructed from eq (3.3) with the following corrections:

1. Add to $\alpha_{oi,j}$ and $S_{i,j}$ the quantities $\Delta_{i,j}$ and $\epsilon_{i,j}$ respectively which are the results when doing the line-integral along the external boundary of the cell.

2. Let $\lambda = \delta = S = 0$ for cells outside the boundary.

As an example we consider the mesh point shown in Fig 3.

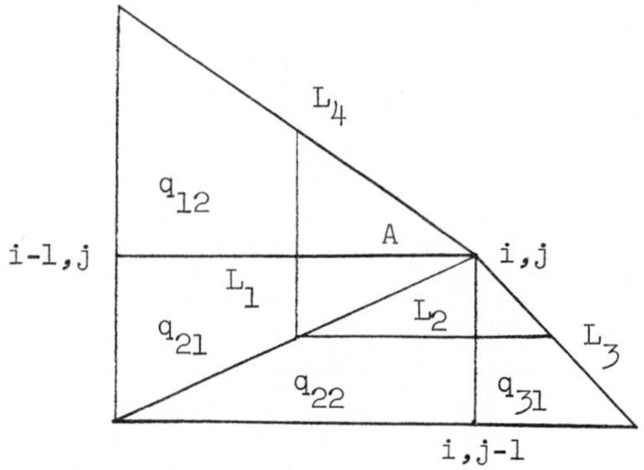

Fig 3. A boundary mesh point.

The integration along L_3 and L_4 results in:

$$\Delta_{i,j} = F_{ij}\ (\alpha) \qquad (3.6)$$

$$\epsilon_{i,j} = F_{ij}\ (\gamma)$$

with the function F defined by:

$$F_{ij}(X) = (X\,\frac{\lambda}{\beta})_{q12}\ \tfrac{1}{2}(h^2_{i-1} + k^2_{j})^{\frac{1}{2}} + (X\,\frac{\lambda}{\beta})_{q31}\ \tfrac{1}{2}(h^2_{i} + k^2_{j-1})^{\frac{1}{2}} \qquad (3.7)$$

Here we have assumed $\beta \neq 0$. When the temperature is known $(\beta = 0)$ the construction of an equation is of course not necessary.

Let N be the number of mesh points representing the unknown temperatures. Then the discrete representation of eqs (2.1) and (2.2) in space may be written:

$$D(\vec{T}) \frac{d}{dt} \vec{T}(t) = A(\vec{T}) \vec{T} + \vec{S}(\vec{T}). \tag{3.8}$$

with the matrices and vectors of order N. The matrix D is diagonal and A is pentadiagonal and symmetrical with a negative diagonal and non-negative offdiagonal elements.

The discretization method which we have used is called the box integration method and was first described by Varga in ref (1). A detailed analysis of the truncation error can be found in ref (3).

From a practical point of view this method is very attractive because the symmetrical property of A neither depends on the geometry of the domain nor of the boundary conditions.

4. TIME INTEGRATION

Four methods for solution of the semi-discrete equation (3.8) will be presented. Two of them are well known from the literature, namely the Alternating Direction Implicit Peacemann Rachford (ADI) method and the Local One Dimensional (LOD) method. The last two methods we think are not so well known. We shall call them the Triangular Split (TS) method and the Frequencial Triangular Split (FTS) method.

4.1 Alternating Direction Implicit

We consider the Peacemann-Rachford method described in ref (4).

The matrix A is written:

$$A = H + V \tag{4.1}$$

where H and V represent the discretization of the operator $\frac{\partial}{\partial x}(\lambda\frac{\partial}{\partial x})$ and $\frac{\partial}{\partial y}(\lambda\frac{\partial}{\partial y})$ respectively together with the boundary conditions.

The ADI method is then written:

$$\begin{aligned}(I - h_n D_n^{-1} H_n)\, \vec{T}_{n+\frac{1}{2}} &= (I + h_n D_n^{-1} V_n)\vec{T}_n \\ &\quad + h_n D_n^{-1} \vec{S}_n \\ (I - h_n D_n^{-1} V_n)\, \vec{T}_{n+1} &= (I + h_n D_n^{-1} H_n)\vec{T}_{n+\frac{1}{2}} \\ &\quad + h_n D_n^{-1} \vec{S}_n\end{aligned} \tag{4.2}$$

Consider the matrix differential equation of order N:

$$\frac{d\vec{\theta}}{dt} = B\,\vec{\theta} + \vec{R} \tag{4.5}$$

Let B be splitted into:

$$B = L + U \tag{4.6}$$

where L and U are the lower and the upper triangular part of B, with

$$\text{diag}\ (L) = \text{diag}\ (U) = \tfrac{1}{2}\ \text{diag}\ (B) \tag{4.7}$$

The eq (4.5) is now integrated from the time t_n to t_{n+1} in the following two steps:

$$\begin{aligned} (I - h_n U_n)\ \vec{\theta}_{n+\frac{1}{2}} &= (I + h_n L_n)\vec{\theta}_n + h_n\vec{R}_n \\ (I - h_n L_n)\ \vec{\theta}_{n+1} &= (I + h_n U_n)\vec{\theta}_{n+\frac{1}{2}} + h_n\vec{R}_n \end{aligned} \tag{4.8}$$

The usefulness of the method depends on the stability and the truncation error. We will not discuss these things here. However, the reader may find it in ref (2).

In eqs (4.8) we see that when the matrix B and the vector $\vec{R}$ are time dependent, they should be evaluated at time t_n in order to calculate $\vec{\theta}_{n+1}$.

When application of the method given by eqs (4.8) to the heat equation, we set:

$$\begin{aligned} B &= D^{-1}A \\ \vec{R} &= D^{-1}\vec{S} \end{aligned} \tag{4.9}$$

then

$$\vec{\theta} = \vec{T} \tag{4.10}$$

4.4 Frequencial Triangular Split

As known from the literature and is easily seen, is that the solution of the matrix equation (3.8) with constant coefficients have an exponential behaviour. Therefore, let

$$\vec{T} = e^{wt} \cdot \vec{\theta}(t) \tag{4.11}$$

where w=w(t) is a diagonal matrix of same order as $\vec{T}$ named the frequence matrix.

Substitution of eq (4.11) into eq (3.8) then results in the following equation for $\vec{\theta}$:

$$\frac{d\vec{\theta}}{dt} = e^{-wt}\,(D^{-1}A - w)\,e^{wt}\vec{\theta} + D^{-1}\vec{S} \tag{4.12}$$

By application of the TS method, the equation is solved according to the following scheme:

$$\begin{aligned}
&[I - h_n e^{-w_n h_n}(U_n - w_n)\,e^{w_n h_n}]\,\vec{\theta}_{n+\frac{1}{2}} = \\
&[I + h_n e^{-w_n h_n} L_n e^{w_n h_n}]\,\vec{\theta}_n + h_n e^{-w_n h_n} D_n^{-1}\vec{S}_n \\
&[I - h_n e^{-w_n h_n}(L_n - w_n) e^{w_n h_n}]\,\vec{\theta}_{n+1} = \\
&[I + h_n e^{-w_n h_n} U_n e^{w_n h_n}]\,\vec{\theta}_{n+\frac{1}{2}} + h_n e^{-w_n h_n} D_n^{-1}\vec{S}_n
\end{aligned} \tag{4.13}$$

In order to obtain the temperature we shall select the frequence matrix so that:

$$\begin{aligned}
&\vec{T}_n = \vec{\theta}_n \\
&\vec{T}_{n+\frac{1}{2}} = e^{w_n h_n}\,\vec{\theta}_{n+\frac{1}{2}} \\
&\vec{T}_{n+1} = e^{2w_n h_n}\,\vec{\theta}_{n+1}
\end{aligned} \tag{4.14}$$

Substitution into eq (4.13) then yields:

$$\begin{aligned}
&[\,I - h_n(U_n - w_n)\,]\,\vec{T}_{n+\frac{1}{2}} = \\
&[I + h_n L_n]\,e^{w_n h_n}\,\vec{T}_n + h_n D_n^{-1}\vec{S}_n \\
&[I - h_n(L_n - w_n)]e^{-w_n h_n}\,\vec{T}_{n+1} = \\
&[\,I + h_n U_n]\;\vec{T}_{n+\frac{1}{2}} + h_n D_n^{-1}\,\vec{S}_n
\end{aligned} \tag{4.15}$$

Because of the triangular properties of the left hand matrices, we see that the inversions are easily performed.

where 2 h_n (n=0,1,2, ...) is the stepsize in time and $\vec{T}_n = \vec{T}(t_n)$. We have assumed that the temperature dependent matrices to be used for calculation of $\vec{T}_{n+1}$ are based on $\vec{T}_n$:

$$\begin{aligned} H_n &= H(\vec{T}_n) \\ V_n &= V(\vec{T}_n) \\ D_n &= D(\vec{T}_n) \end{aligned} \tag{4.3}$$

We know from ref (4) that the method is unconditionally stable at least when the domain is rectangular and the problem is linear with constant coefficients and a constant stepsize. A theoretical analysis of the nonlinear problem is not known to exist to the authors. However, numerical experiments indicate some stability limits.

4.2 Locally One Dimensional

The matrix A is again written as eq (4.1). However, instead of eq (4.2) we have the following algorithm:

$$\begin{aligned} (I - h_n D_n^{-1} H_n)\, \vec{T}_{n+\frac{1}{2}} &= (I + h_n D_n^{-1} H_n) \vec{T}_n + h_n D_n^{-1} \vec{S}_n \\ (I - h_n D_n^{-1} V_n) \vec{T}_{n+1} &= (I + h_n D_n^{-1} V_n) \vec{T}_{n+\frac{1}{2}} + h_n D_n^{-1} \vec{S}_n \end{aligned} \tag{4.4}$$

An analysis of the LOD method is also given in (4), and it is proved that the method is unconditionally stable under the same conditions as for the ADI method.

Later we shall give some numerical results which indicates that the method is also useful to non-linear problem.

4.3 Triangular Split

Certainly, both the ADI and the LOD methods represent splitting methods. The matrix A is splitted in such a way that in the case of 2-dimensional problems the solution of the matrix equation for each half time step is easily performed because of the tri-diagonal property.

We shall now split the matrix into an upper and a lower matrix. This means that the solution is very easily obtained. Only a substitution procedure is required for each step.

Both the schemes (4.15) and (4.8) are unconditionally stable applied to the heat conduction problem with constant stepsize and coefficients. In ref (2) Reed and Hansen present a detailed analysis of the method.

The function $\vec{\theta}$ in eq (4.11) may be looked on as a correction to the constant cofficient problem. When applied to such problems it is given by the initial condition. The frequence matrix is then given by the eigenvalues of $D^{-1}A$. Now, in a realistic problem the frequences are not known. However, we may estimate them by:

$$w_{in} = \frac{1}{2\ h_n} \ln \left(\frac{T_{in}}{T_{in-1}}\right) \quad ; \quad i = 1,2,\ldots N \tag{4.16}$$

When obtained a value for $\vec{T}_{n+1}$ based on $\left\{w_{in}\right\}$ an "error" in the frequences may be calculated from:

$$e_{n_i} = \frac{1}{2h_n} \ln \left(\frac{T_{i\ n+1}}{T_{in}}\right) - \frac{1}{2\ h_n} \ln \left(\frac{T_{in}}{T_{in-1}}\right) \tag{4.17}$$

which gives the possibility of changing the stepsize in time and keeping full control of the error during the integration.

5. APPLICATIONS

The methods have been applied to two specific problems. The first is a linear problem with constant coefficients and Dirichlets type of boundary conditions and the second is a problem with temperature dependent coefficients and boundary conditions containing both conduction and radiation. In the last problem we have also included an internal radiation zone.

5.1 Constant Coefficients

Let the domain of definition in space be:

$$\tau = \left\{(x,y);\ \ x\epsilon[0,1],\ \ y\epsilon[0,1]\right\}$$

The equation is then written

$$\frac{\partial}{\partial t} T(\vec{r},t) = \sigma \cdot \nabla^2 T(\vec{r},t); \tag{5.1}$$
$$\vec{r} \in \tau,\ t \geq 0$$

with the boundary condition

$$T(\vec{r},t) = 0 \; ; \; \vec{r} \in \partial \tau \tag{5.2}$$

and the initial condition

$$T(\vec{r},0) = \sin(\pi x) \cdot \sin(\pi y) \; ; \; \vec{r} \in \tau \tag{5.3}$$

The analytical solution is given by:

$$T_A(\vec{r},t) = \exp(-2\pi^2 \sigma t) \sin(\pi x) \cdot \sin(\pi y) \tag{5.4}$$

For the constant σ we select the approximate value of steel, 10^{-4} m^2/s.

On the domain is imposed 13 mesh points equal spaced, in each direction.

In the Tables I to IV the error defined as

$$E = \max_{i,j} \left| \frac{T_{i,j} - T_{Ai,j}}{T_{Ai,j}} \right|$$

is given for each of the methods for different time steps and at different times.

Table I

The A.D.I.

h(s) \ t(s)	500	1000
1.0	$4.9 \cdot 10^{-3}$	$3.7 \cdot 10^{-3}$
10.0	$6.8 \cdot 10^{-2}$	$4.6 \cdot 10^{-2}$
20.	0.15	$8.4 \cdot 10^{-2}$
30.	0.12	0.11
40.	0.29	0.11
50.	0.36	0.14
60.	0.40	0.17
70.	8.8	120
80.	210	$5 \cdot 10^{4}$
100.	10^{6}	10^{10}

Table II

The L.O.D.

h(s) \ t(s)	500	1000	1200	2400	3600	4800
1.0	$1.5 \cdot 10^{-3}$	$1.1 \cdot 10^{-3}$				
10.	$1.5 \cdot 10^{-3}$	$1.1 \cdot 10^{-3}$				
20.	$1.5 \cdot 10^{-3}$	$1.0 \cdot 10^{-3}$				
30.	$1.4 \cdot 10^{-3}$	$1.0 \cdot 10^{-3}$				
40.	$1.5 \cdot 10^{-3}$	$1.0 \cdot 10^{-3}$				
50.	$1.5 \cdot 10^{-3}$	$1.0 \cdot 10^{-3}$				
60.	$1.5 \cdot 10^{-3}$	$1.0 \cdot 10^{-3}$				
70.	$1.4 \cdot 10^{-3}$	$9.6 \cdot 10^{-4}$				
80.	$1.3 \cdot 10^{-3}$	$9.2 \cdot 10^{-4}$				
90.	$1.0 \cdot 10^{-3}$	$9.0 \cdot 10^{-4}$				
100.	$9.1 \cdot 10^{-4}$	$7.1 \cdot 10^{-4}$	$1.3 \cdot 10^{-4}$	$2.0 \cdot 10^{-5}$	$2.6 \cdot 10^{-6}$	
200.			$2.1 \cdot 10^{-3}$	$3.8 \cdot 10^{-4}$	$5.3 \cdot 10^{-5}$	
300.			$5.7 \cdot 10^{-3}$	$1.0 \cdot 10^{-4}$	$1.4 \cdot 10^{-4}$	
400.*			$6.6 \cdot 10^{-3}$	$5.1 \cdot 10^{-4}$		$3.0 \cdot 10^{-5}$
600.			$2.7 \cdot 10^{-2}$	$4.4 \cdot 10^{-3}$	$5.2 \cdot 10^{-4}$	
1200.				$2.0 \cdot 10^{-3}$		$3.2 \cdot 10^{-5}$
1800.					0.08	

* The time is taken at t = 1600, 3200 and 4800 seconds

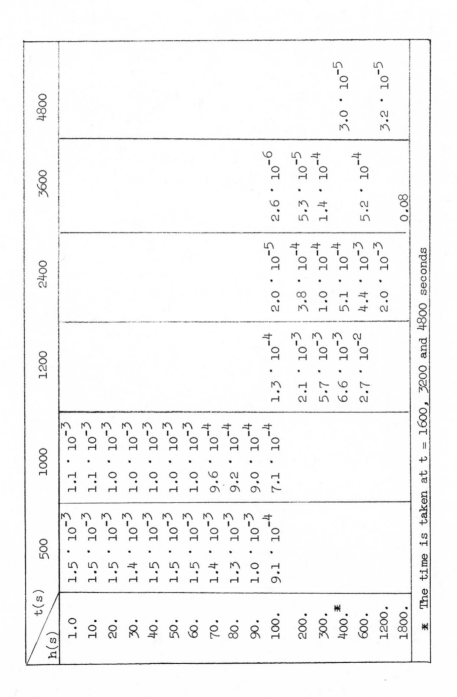

Table III

The T.S.

h(s) \ t(s)	500	1000
1.0	$1.5 \cdot 10^{-3}$	$1.1 \cdot 10^{-3}$
10.0	$2.7 \cdot 10^{-3}$	$1.9 \cdot 10^{-3}$
20.	$6.2 \cdot 10^{-3}$	$4.3 \cdot 10^{-3}$
30.	$1.2 \cdot 10^{-2}$	$8.2 \cdot 10^{-3}$
40.	$2.0 \cdot 10^{-2}$	$1.3 \cdot 10^{-2}$
50.	$3.0 \cdot 10^{-2}$	$2.3 \cdot 10^{-2}$
60.	$4.1 \cdot 10^{-2}$	$3.4 \cdot 10^{-2}$
70.	$5.3 \cdot 10^{-2}$	$4.0 \cdot 10^{-2}$
80.	$6.8 \cdot 10^{-2}$	$5.7 \cdot 10^{-2}$
90.	$8.4 \cdot 10^{-2}$	$6.8 \cdot 10^{-2}$
100.	0.10	$7.6 \cdot 10^{-2}$

Table IV

The FTS

h(s) \ t(s)	500	1000	1200	2400	3600	4800
1.0	$1.5 \cdot 10^{-3}$	$1.1 \cdot 10^{-3}$				
10.	$1.5 \cdot 10^{-3}$	$1.1 \cdot 10^{-3}$				
20.	$1.5 \cdot 10^{-3}$	$1.0 \cdot 10^{-3}$				
30.	$1.5 \cdot 10^{-3}$	$1.0 \cdot 10^{-3}$				
40.	$1.5 \cdot 10^{-3}$	$9.5 \cdot 10^{-4}$				
50.	$1.5 \cdot 10^{-3}$	$1.1 \cdot 10^{-3}$				
60.	$1.5 \cdot 10^{-3}$	$1.1 \cdot 10^{-3}$				
70.	$1.5 \cdot 10^{-3}$	$1.1 \cdot 10^{-3}$				
80.	$1.5 \cdot 10^{-3}$	$1.0 \cdot 10^{-3}$				
100.	$1.5 \cdot 10^{-3}$	$1.0 \cdot 10^{-3}$	$9.9 \cdot 10^{-4}$	$1.7 \cdot 10^{-4}$	$2.3 \cdot 10^{-5}$	
200.			$9.2 \cdot 10^{-4}$	$1.6 \cdot 10^{-4}$	$2.3 \cdot 10^{-5}$	
300.			$8.4 \cdot 10^{-4}$	$1.5 \cdot 10^{-4}$	$2.2 \cdot 10^{-5}$	
400.			$5.1 \cdot 10^{-4}$	$4.3 \cdot 10^{-5}$		$2.8 \cdot 10^{-6}$
600.			$8.1 \cdot 10^{-4}$	$1.5 \cdot 10^{-4}$	$2.2 \cdot 10^{-5}$	
1200.				$1.6 \cdot 10^{-4}$		$2.7 \cdot 10^{-6}$
1800.					$2.2 \cdot 10^{-5}$	

* The time is taken at t = 1600, 3200 and 4800 seconds

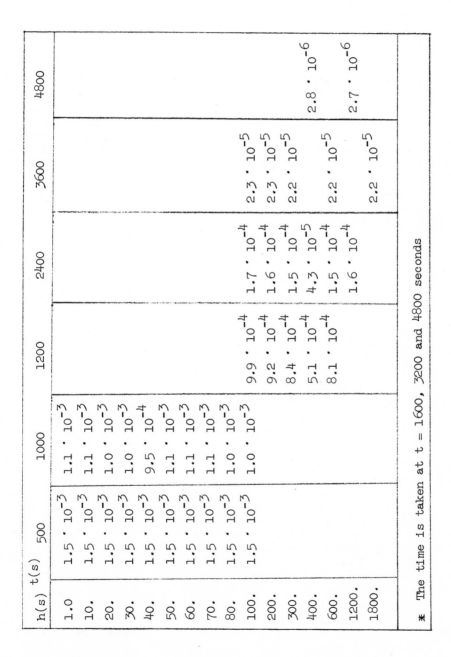

Defining p as:

$$p = \sigma \cdot \Delta t/(\Delta X)^2$$

we get the following conditions on p for each of the methods in order to keep the error under control.

ADI	$p \leq 2$
LOD	$p \leq 40$
TS	$p \leq 6$
FTS	no upper limit

The algorithm of Thomas given in ref (4) was used for solving the matrix equation on each half step for both the ADI and LOD method. Theoretically both the methods are unconditionel stable. The reason why the LOD is about 20 times "better" than the ADI is difficult to answer. However, it seemes that the LOD method has better positive properties. By the sentence "positive properties", we here mean an internal property of an integration scheme:

$$B_1\vec{U}_{n+1} = B_2\vec{U}_n+\vec{C}; \; n=0,1,\ldots$$

so that

$$\text{sign}\,(\vec{U}_{n+1}) = \text{sign}\,(\vec{U}_n) \; ; \; n \geq n0$$

where n0 is a positive integer.

The TS method is "better" than the ADI method by a factor of 3. The exponential transformation done by introducing the frequence matrix in the TS method seems to be very powerful. The error is almost independent of the size of the time step.

As expected, for doing one step of integration both the ADI, LOD and the TS methods are about 2 times faster than the FTS. This is owing to the exponential term in the last method. However, the FTS method was the most accurate one.

5.2 Variable Coefficients

The problem is to calculate the temperature distribution in a rectangular cross section of an ingot of steel contained in the mould during the period between teeming and stripping. Thermal conductivity, specific heat and density are known functions of the temperature. The problem is really a two phase problem. And in this example we have included the latent heat into the specific heat function. The equation for this problem is given by eq (2.1)

During the first 300 seconds after the teeming we assume that the steel is in contact with the mould. Therefore, during this period we have conduction from steel to mould. After this first period we assume that an air gap of 0.01 m between ingot and mould is developed. The heat transfer mechanism from the ingot to the mould is now radiation.

The boundary conditions include both conduction and radiation and are therefore a combination of type (iii) and (v) in chapter 2.

An initial temperature distribution is given.

On the domain we impose 25 mesh points in each direction, so that the step sizes in each direction for the ingot, the air gap (if any) and the mould are 0.05 m, 0.01 m and 0.06 m respectively. Due to the symmetrical property of the problem, we only calculate the temperature distribution in the first quadrant. The actual calculation is therefore performed with 169 mesh points.

In Table V we have given the error for different time step sizes at time t = 1000 s.

Table V

h(s)	10.	20.	50.	100.	150.
F.T.S.	1.0	1.4	2.1	3.3	5.0
A.D.I.	7.	11.	35.	58.	78.
T.S.	6.	10.	37.	55.	65.
L.O.D.	1.5	1.7	6.1	11.5	25.

No analytical solution is known. Therefore the error is defined as:

$$\epsilon = \max_{i,j} |T_{ij} - T_{cij}|$$

where T_c denotes the assumed correct answer obtained by a refined calculation. This temperature distribution has also been compared to experimental measurements with a very good agreement.

For this specific problem we obtained the following stability limits on the time step Δt:

ADI	Δt = 200 s
LOD	Δt = 1000 s
TS	Δt = 200 s
FTS	Δt = 1000 s

From a stability point of view, the best methods are the LOD and the FTS. However, looking at the error we see that the FTS method is the most accurate. Therefore, if we require an answer within 1% error which is reasonable in most engineering calculations, the FTS method is more than 5 times faster than the other methods. The execution time in order to simulate 3 hours real time was 4 s.

REFERENCES

(1) Richard S. Varga: "Matrix Iterative Analysis". Prentice-Hall, Inc. (1962).

(2) William H. Reed, K.F. Hansen: "Finite Difference Techniques for the Solution of the Reactor Kinetics Equations". MIT-3903-2 MITNE-100 (1969).

(3) Eugene L. Wachspress: "Iterative Solution of Elliptic Systems". Prentice-Hall, Inc. (1966).

(4) A.R. Mitchell: "Computational Methods in Partial Differential Equations". John Wiley & Sons (1969).

(5) Walter Rudin: "Principles of Mathematical Analysis". McGraw-Hill Book Company (1964).

FINITE ELEMENT COMPUTER PROGTAMS FOR HEAT CONDUCTION PROBLEMS

E. Spreeuw

Stichting Reactor Centrum Nederland

The finite element method in its present form has been introduced by civil engineers. The most important reason for the development of it existed from the desire to solve difficult problems from structural mechanics in a relatively simple way. From the viewpoint of simplicity of formulation it is surprising that the method has been developped for structural analysis, because the application to the solution of many other partial differential equations, e.g. like exist for temperature distributions in solids, is definitely more easy to perform. This is true since in structural analysis one deals with displacement (vector) fields influenced by contractions which implies a complicated relation between stresses and displacements. Correspondingly in thermal analysis one is interested in temperature (scalar) distributions while the relation between heat fluxes and temperatures is given by the simple expression of Fourier's law of heat conduction.
Steady state temperature distributions in solids may also be solved applying Ritz method and making use of the connection established between this method and the finite element approach allows to ensure convergence as far as completeness is achieved (Ref. |2|). The classical method of Galerkin in fact means an extension of Ritz method and it may be used for the finite element formulation of a large class of physical problems. Galerkin's method has been applied here to derive a functional expression for thermal analyses in order to elucidate this broad applicability. The realization in two computer programs will be discussed.

1. THE GENERAL EQUATION OF BALANCE.

For every transferable property E per unit mass, the general equation of balance per unit volume reads:

$$\frac{\partial}{\partial t}(\rho E) + \frac{\partial}{\partial x_j}(\rho E \dot{v}_j) + \frac{\partial q_j}{\partial x_j} - S = 0 \qquad (1)$$

Here the Eulerian representation has been used, in which the phenomenon under consideration has been investigated in relation to a fixed coordinate system.

The first term of the equation involves the time rate of property E within the unit control volume while the second and third terms represent the net rate of E out of the unit volume through the boundaries by convective and contact transport respectively.
The quantity S signifies all contributions to the equations originating from internal or external causes. ρ stands for density while q_j and $\dot{v}_j$ mean the components in the x_j direction of contact transport and convective velocity vectors respectively. Moreover, here the summation convention is used which states that subscripts repeated within a term automatically imply summation over those subscripts.
In general the finite element method can be applied if a vanishing variational expression can be derived from equation (1), the constitutive equations and the relevant boundary and initial conditions. The practical application is hampered however if property E includes velocities. The resulting system of equations will then be nonlinear and the numerical solution will be comprehensive. The variational expression used for the development of most of the existing finite element computer programs means the variation of a functional which satisfies an extremum principle. The most familiar example of such a functional is the system's potential energy used in structural mechanics which satisfies the principal of minimum potential energy. Similarly we have the principle of minimum entropy production in fluid dynamics and Hamilton's principle in mechanics. In agreement with Galerkin's method, also used by Visser (Ref. |1|) a variational expression will be derived for the problem of heat conduction in solids. Also in this particular application the variational expression refers to the existence of a functional with an extreme value. An alternative way to arrive at a starting point for finite element application can be followed after the particular form of the equation(s) of balance of a specific problem are combined with its constitutive equation(s) to obtain a differential equation of the form:

$$Au = f,$$

where A is a linear positive definite matrix differential operator. In this method a functional analytical approach is followed in that a basis is constructed in a Hilbert space and the functional used is defined by

$$F(u) = |u,u| - 2\,(u,f),$$

with

$$|u,u| = \int_V (Au)^T.u\; dV$$

and

$$(u,f) = \int_V u^T.f\; dV$$

Formulations like this one enable to proof the existence, uniqueness and convergence of solutions (Refs. |2| and |3|).

2. A FUNCTIONAL EXPRESSION FOR THE PROBLEM OF HEAT CONDUCTION.

In the case of heat transfer E equals scalar $c\Theta$ and consequently the equation of heat balance becomes

$$\frac{\partial}{\partial t}(\rho c\Theta) + \frac{\partial}{\partial x_j}(\rho c\Theta\dot{v}_j) + \frac{\partial q_j}{\partial x_j} - Q^o = 0 \qquad (2)$$

Θ meaning temperature, c specific heat and Q^o volumetric heat generation rate. By restricting the investigation to heat conduction in solids the second term of (2) vanishes. On the assumption that ρ and c are independent of temperature the equation becomes:

$$-\frac{\partial q_i}{\partial x_i} + Q^o - \rho c\dot{\Theta} = 0 \qquad (3)$$

On the surface of the solid in question three kinds of boundary conditions can be considered:

1. prescribed temperature $\Theta = \Theta^o$,
2. prescribed heat input $q_i n_i = -\overset{o}{q_i n_i}$, n_i representing the outward unit normal to the boundary surface,
3. heat input proportional to the difference of the temperatures of the structure surface and the surrounding medium $q_i n_i = \alpha(\Theta-\Theta_m)$, α means the coefficient of heat transfer between the medium and the structure.

The surfaces, to which these boundary conditions are applicable, together build A, which is the whole external surface of the structure to be investigated.

Equation (3) and the homogenized boundary conditions

$$q_i n_i + q_i^o n_i = 0 \quad (\text{on } A^q) \quad \text{and}$$

$$q_i n_i - \alpha(\Theta - \Theta_m) = 0 \quad (\text{on } A^\alpha)$$

are multiplied by an arbitrary function $\delta\Theta$ of the coordinates which however vanishes on that particular fraction of the geometry where the temperature is prescribed. $\delta\Theta$ acts as a Langrange multiplier and successive integration over volume V and the applicable surfaces leads to

$$\delta P = - \iiint_V \left\{ \frac{\partial q_i}{\partial x_i} - Q^o + \rho c\dot{\Theta} \right\} \delta\Theta dV - \iint_A q_i n_i \delta\Theta dA - \iint_{A^q} q_i^o n_i \delta\Theta dA +$$

$$+ \iint_{A^\alpha} \alpha(\Theta - \Theta_m)\delta\Theta dA = 0$$

According to Gauss' divergence theorem

$$\iint_A q_i n_i \delta\Theta dA = \iiint_V \frac{\partial(q_i \delta\Theta)}{\partial x_i} dV$$

Choosing for $\delta\Theta$ differentiable variations of the temperature distribution Θ makes variational calculus applicable, so that:

$$\frac{\partial(q_i\delta\Theta)}{\partial x_i} = \frac{\partial q_i}{\partial x_i}\delta\Theta + q_i \frac{\partial(\delta\Theta)}{\partial x_i} = \frac{\partial q_i}{\partial x_i}\delta\Theta + q_i \delta\left(\frac{\partial\Theta}{\partial x_i}\right)$$

Hence:

$$\delta P = -\iiint_V \{q_i \delta(\frac{\partial\Theta}{\partial x_i}) + (Q^o - \rho c\dot{\Theta})\delta\Theta\} dV - \iint_{A^q} q_i^o n_i \delta\Theta dA +$$

$$+ \iint_{A^\alpha} \alpha(\Theta - \Theta_m)\delta\Theta dA = 0$$

For linear problems, the heat conductivity, specific heat and heat transfer coefficient are all independent of temperature. The constitutive equation embodying the relation between heat flow and temperature is Fourier's law of heat conduction:

$$q_i = -\lambda_{ij} \frac{\partial\Theta}{\partial x_j} \qquad i = 1,2,3$$

By means of statistical, mechanical considerations Onsager arrived at his postulate stating that the heat conductivity matrix must be a symmetric one:

$$\lambda_{ij} = \lambda_{ji} \qquad i,j = 1,2,3$$

Moreover, when only studying isotropic materials

$$\lambda_{ij} = \lambda\delta_{ij} \qquad i,j = 1,2,3$$

λ represents the direction independent coefficient of heat conductivity while Kronecker's delta δ_{ij} is given by:

$$\delta_{ij} = \left| \begin{array}{l} 1 \text{ when } i = j \\ 0 \text{ when } i \neq j \end{array} \right.$$

Now δP can be replaced by:

$$\delta P = \iiint_V \{\lambda \frac{\partial\Theta}{\partial x_i} \delta(\frac{\partial\Theta}{\partial x_i}) - (Q^o - \rho c\dot{\Theta})\delta\Theta\} dV - \iint_{A^q} q_i^o n_i \delta\Theta dA +$$

$$+ \iint_{A^\alpha} \alpha(\Theta - \Theta_m)\delta\Theta dA = 0 \qquad (4)$$

The approach followed to arrive at this expression may be used for a large class of problems satisfying equation (1).
Because there is no integration over time, $\dot{\Theta}$ can be treated as a fixed quantity and as no variation of $\dot{\Theta}$ exists.

$$\delta P = \delta\{\iiint_V \left(\frac{\lambda}{2} \frac{\partial \Theta}{\partial x_i} \frac{\partial \Theta}{\partial x_i} - Q^o \Theta + \rho c \dot{\Theta} \Theta \right) dV - \iint_{A^q} q_i^o n_i \Theta dA +$$

$$+ \iint_A \frac{\alpha}{2} (\Theta - \Theta_m)^2 dA\} = 0$$

So if we call

$$P = \iiint_V \left(\frac{\lambda}{2} \frac{\partial \Theta}{\partial x_i} \frac{\partial \Theta}{\partial x_i} - Q^o \Theta + \rho c \dot{\Theta} \Theta \right) dV - \iint_{A^q} q_i^o n_i \Theta dA +$$

$$+ \iint_{A^\alpha} \frac{\alpha}{2} (\Theta - \Theta_m)^2 dA \qquad (5)$$

then equation (4) can be interpreted as a necessary condition for P to reach an extreme value with respect to admissible variations of the temperature distribution Θ.
P is known as the functional of the problem of heat conduction.
Varying temperature we find that the distribution Θ that gives an extremum of this functional must satisfy the Euler-Lagrange equation:

$$\lambda \frac{\partial^2 \Theta}{\partial x_i \partial x_i} + Q^o - \rho c \dot{\Theta} = 0 \qquad (6)$$

This relation is indeed found after substitution of Fourier's law of heat conduction in equation (3).
When restricted to stationary conditions equation (6) can be reduced to a form

$$Au = f \qquad (7)$$

with u and f elements of a Hilbert space H.
A is an operator transforming elements from its area of definition $D(A) \subset H$ in H.
Operator A satisfies the following requirements:
A is linear, this means A as well as D(A) are linear
A is symmetric, so $(Au,v)=(u,Av)$ with (u,v) is the inner product defined in H
A is positive definite, this means $(Au,u) \geqq \gamma(u,u)$ if $\gamma>0$.

We will now proof that if equation (6) has a solution, then it will minimize the functional

$$F(u) = (Au,u) - 2(u,f) \qquad (8)$$

It is supposed that F(u) is only defined on D(A), so if F(u) attains a minimum value for an element u_o, this element will automatically be present in D(A).
We first define the energyproduct $|u,v|_A = (Au,v)$ and the energynorm $||u||_A = |u,u|_A$

If $u=u_o$ is a solution of equation (7) then $Au_o=f$ and

$$\begin{aligned} F(u) &= (Au,u) - 2(u,Au_o) \\ &= |u,u|_A - 2|u,u_o|_A \\ &= ||u-u_o||_A^2 - ||u_o||_A^2 \end{aligned}$$

From this we see that F(u) becomes minimal if $u=u_o$.
We also have to prove the reverse, implying that if F(u) attains a minimum for an element $u=u_o$, this element will be the solution of equation (7).
D(A) is a linear space, so if $\eta \in D(A)$ and $\lambda \in R^1$ then also $\{u_o + \lambda\eta\} \in D(A)$. $F(u_o)$ is minimal, so:

$$F(u_o + \lambda\eta) - F(u_o) \geq 0 \text{ for every } \eta \in D(A) \text{ and } \lambda \in R^1$$

This means:

$$(A(u_o+\lambda\eta),\, u_o+\lambda\eta) - 2(u_o+\lambda\eta,f) - (Au_o,u_o) + 2(u_o,f) \geq 0$$

Thus:

$$2\lambda(Au_o-f,\eta) + \lambda^2(A\eta,\eta) \geq 0$$

From discriminant ≤ 0 follows

$$(Au_o-f,\eta)^2 \leq 0$$

So

$$(Au_o-f,\eta) = 0$$

This holds for every $\eta \in D(A)$.

A wellknown theorem from functional analysis reads:
If an element from a Hilbert space H is perpendicular to all elements from a closed collection $D \subset H$, then this element is the zero element from H.
Therefore $Au_o=f$.

When neglecting a factor 2 the expression F(Θ) derived from equation (6) and its boundary conditions becomes identical to P (Eq. (5)).

3. REPRESENTATION BY FINITE ELEMENTS.

For the finite element formulation of heat conduction problems, P will no longer be expressed as a functional of the continuous temperature distribution, but in terms of a number of discrete values. These values consist of nodal point temperatures and/or temperature derivatives, called the unknowns or the degrees of freedom. The solution of the problem is now obtained from the vanishing of δP for all variations of the degrees of freedom. In the stationary case this leads to a set of linear algebraïc equations while in the transient case one obtains a system of first order linear differential equations. The number is equal to the number of unknowns.
For a finite element calculation we proceed as follows:

1. The model of the structure to be investigated is divided in a finite number of "finite elements".
2. The elements are assumed to be interconnected at a discrete number of nodal points situated at their boundaries. The temperatures and/or temperature derivatives at these nodal points are introduced as the degrees of freedom.
3. The temperature distribution is approximated by an interpolating polynomial, called the "temperature function" uniquely defining the state of temperature within each "finite element" in terms of coefficients which depend on its nodal values.

   The temperature function has to fulfil the following consistency conditions:

   a. At least homogeneous and linearly varying temperature distributions are described. This is sometimes called the completeness criteria.
   b. Temperatures along elemental boundaries must be uniquely defined when related to the distributions valid inside each of the adjacent elements; this is the compatibility condition.

   The first of these conditions is necessary, the second one is only sufficient. In case this one is not met the convergence of the solution has to be traced since convergence will not be assured automatically.
   Another useful condition that has to be mentioned implies invariance of the finite element solution, i.e. independence of the position of the external (global) reference system.
   The temperature functions should therefore be independent of the orientation of the element with respect to the coordinate

axes to which they are referred. This invariance condition is automatically satisfied if the temperature function is a complete polynomial expression. If this independence does not exist, all elemental contributions to the resulting system of equations have to be related to a global coordinate system. This leads to the introduction of transformation matrices, which however will not be discussed here.

Substitution of the nodal point coordinates leads to a set of equations from which the coefficients of the temperature function can be derived in terms of the unknowns. This relation is given by the so-called combination matrix. Existance of this matrix requires the number of terms describing the temperature profile along a boundary to be at least equal to the number of unknowns along that boundary.

4. The integrands of equation (4) can be calculated and in order to get an expression in terms of the degrees of freedom the vector of coefficients is replaced by the product of the combination matrix and the vector of unknowns.
5. Now the integrations of equation (4) can be evaluated and, for every element the contribution to δP is found in terms of variations of the unknowns.

For a structure idealized by only one element, the vanishing of δP for each variation of the unknowns leads to a number of equations equal to the number of elemental unknowns. For more complex problems the unknowns have to be numbered singly over the whole structure. With a total number of n unknowns the solution of equation (4) leads to a set of n equations.
For the set-up of these equations the contribution of δP per element must be added successively in the coefficient matrix and in the right hand side of the whole set, i.e. matrix superposition. The indices of the coefficients, to which the δP contribution has to be added, depend on the relation between the elemental unknown and the whole structure unknown numbering. This can be formulated as follows: two unknowns of an element being denoted by indices i and j and these unknowns being indicated within the system of whole structure unknowns by subscripts k and m, the matrix element of a single finite element matrix indicated by i, j must be added to the matrix element of the corresponding whole structure matrix indicated by k, m.

To satisfy the boundary condition of prescribed temperatures, the values of the unknowns Θ_j^o concerned are substituted while the corresponding equations, i.e. the corresponding rows of the coefficients matrix and the components of the right hand term are neglected.

4. THE FIESTA-2 COMPUTER CODE.

The afore mentioned procedure has been applied in developping the Fiesta-2 code. Fiesta stands for Finite Element Stress and Temperature Analysis and the code allows computation of steady-state temperatures and stresses in two- and threedimensional solids. Using Fiesta-2 axisymmetric geometries are idealized by solid of revolution elements with either a rectangular or a triangular cross section. The second type of these ring elements is used only to account for curved or conical boundaries. The cross sections of all ring elements have one side parallel and one side perpendicular to the axis of revolution; thus the cross section of each axisymmetric solid is mapped by an arrangement of parallel and perpendicular mesh lines and only radial and axial step sizes have to be given in the input data deck in order to allow the idealization of these solids. A ring sector element is present for those cases where only a part of the circumference of an axisymmetric body has to be investigated or where boundary conditions, loads or material properties vary along the circumferential direction of the structure. A ring sector plate element is available for two-dimensional problems. Obviously for sector elements also the sector angles have to be submitted in the input data.

Two- and threedimensional geometries described by rectangular coordinates can be idealized by rectangular plate and parallelepiped elements respectively. If appropriate to approximate oblique or curved sides, these elements can be supplemented by triangular plate and three-sided straight prism elements. In agreement with the idealization of geometries described by cylindrical coordinates the idealization can be performed by mapping the two- or three-dimensional model with a pattern of parallel and perpendicular mesh lines, while only the step sizes in each direction have to be given in the input data.
For all of the geometries the grid point and element numbering as well as the element connection and topographical data are generated during execution of the computor program.
The type of mesh generator mentioned means an important simplification when compared to other computor codes where the coordinate values of all grid point and the connection data of all elements have to be submitted.

All elements have grid points located at each angular point. For thermal calculations the degrees of freedom per grid point are represented by the nodal temperatures, while for stress calculations the nodal displacement components in the directions of the appropriate coordinate axes act as unknowns. In order to obtain improved accuracy when axisymmetric problems have to be analyzed, the number of elements is supplemented with two elements with degrees of freedom not only represented by the nodal temperatures but also by their derivatives in axial and radial directions.

Since also nodal derivatives are used to approximate the temperature distribution inside these elements, the type of interpolation is called Hermitean.

The restricted number of data necessary to describe the elemental geometries enabled to perform all matrix multiplications analytically. The computor code uses the results of these multiplications which caused an important reduction of error accumulation and consequently led to improved accuracy of heat conduction matrices and load vectors.

4.1 Finite elements with Hermitean interpolation

The derivation of finite element matrices and vectors will now be discussed on the basis of the Hermitean ring elements implemented in Fiesta-2 (Fig. 1).
Since the number of coefficients contained in an interpolating polynomial has to equal the number of degrees of freedom of the element, a twelve term polynomial has to be chosen for the Ring-12 element. In fact four unknowns are located at each boundary of this element and therefore compatibility requires the interpolating polynomial to enable the approximation of the temperature profile along each boundary by a third order polynomial expression.
A distribution also meeting the conditions of invariance and completeness is:

$$\Theta = a_0 + a_1 z + a_2 z^2 + a_3 z^3 + a_4 r + a_5 zr + a_6 z^2 r + a_7 z^3 r + a_8 r^2 + a_9 zr^2 + a_{10} r^3 + a_{11} zr^3 \quad (9)$$

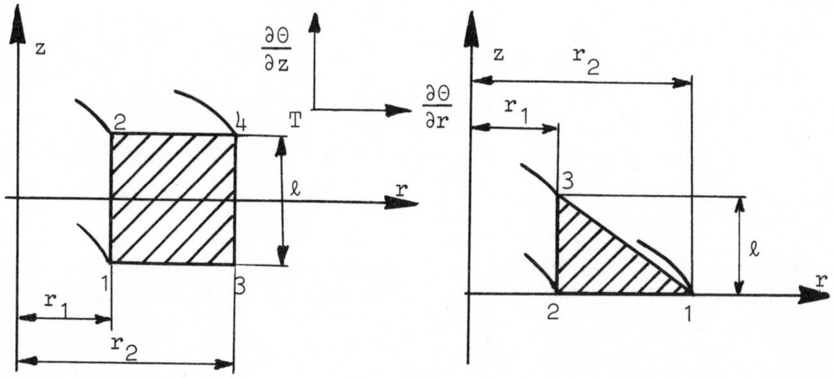

Rectangular cross section, Ring-12 Triangular cross section, Ring-9

Fig. 1: Two Hermitean ring elements

Similarly the temperature inside the Ring-9 element is approximated by:

$$\Theta = b_0 + b_1 z + b_2 r + b_3 z^2 + b_4 zr + b_5 r^2 + b_6 z^3 + b_7 r^3 + \\ + b_8 zr(z+r) \qquad (10)$$

The irregular form of the last term of this expression was chosen not only to fulfil the desire to interpolate symmetrically in axial and radial directions, but also to satisfy the condition of invariance.

Since the elements concerned imply axisymmetry it is obvious to transform equation (4) to cylindrical coordinates. Integration over the angular direction and neglect of a 2Π factor leads to:

$$\delta P_e = \iint_{A^c} \{\lambda \frac{\partial\Theta}{\partial x_i} \delta(\frac{\partial\Theta}{\partial x_i}) - (Q^o - \rho c\dot{\Theta})\delta\Theta\} r dA - \int_{\ell^q} q_i^o n_i \delta\Theta r d\ell + \\ + \int_{\ell^\alpha} \alpha(\Theta - \Theta_m)\delta\Theta r d\ell \qquad (11)$$

A^c representing the surface of the elemental cross-section, ℓ^q and ℓ^α the borders of this area to which boundary conditions apply.

For the Ring-12 geometry we find:

$$\delta P_e = \int_{r_1}^{r_2}\int_{-\frac{\ell}{2}}^{\frac{\ell}{2}} \lambda\{\delta(\frac{\partial\Theta}{\partial z}) \frac{\partial\Theta}{\partial z} + \delta(\frac{\partial\Theta}{\partial r}) \frac{\partial\Theta}{\partial r}\} r dz dr + \int_s \alpha\delta\Theta\Theta r ds - \\ - \int_{r_1}^{r_2}\int_{-\frac{\ell}{2}}^{\frac{\ell}{2}} Q^o \delta\Theta r dz dr - \int_s (q^o + \alpha\Theta_m)\delta\Theta r ds + \int_{r_1}^{r_2}\int_{-\frac{\ell}{2}}^{\frac{\ell}{2}} \rho c \delta\Theta\dot{\Theta} r dz dr \qquad (12)$$

s denoting coordinates along the boundaries of the cross-section. The temperature distribution (Eq. (9)) can be written as a vectorial inner product

$$\Theta = (a,\Phi) = a^T\Phi$$

where a and Φ mean vectors in 12-dimensional Euclidean space; a has components a_o to a_{11} and the elements of Φ contain products of z and r powers

$$\Phi^T = (1,z,z^2,z^3,r,zr,z^2r,z^3r,r^2,zr^2,r^3,zr^3)$$

Consequently

$$\frac{\partial\Theta}{\partial z} = a^T\Phi_z \quad \text{and} \quad \frac{\partial\Theta}{\partial r} = a^T\Phi_r \tag{13}$$

with

$$\Phi_z^T = (0,1,2z,3z^2,0,r,2zr,3z^2r,0,r^2,0,r^3)$$

and

$$\Phi_r^T = (0,0,0,0,1,z,z^2,z^3,2r,2zr,3r^2,3zr^2)$$

Thus Eq. (12) can be rewritten as

$$\delta P_e = \int_{r_1}^{r_2}\int_{-\frac{\ell}{2}}^{\frac{\ell}{2}} \lambda\{\delta a^T\Phi_z\Phi_z^T a + \delta a^T\Phi_r\Phi_r^T a\} r dz dr + \int_s \alpha\delta a^T\Phi\Phi^T a r ds -$$

$$- \int_{r_1}^{r_2}\int_{-\frac{\ell}{2}}^{\frac{\ell}{2}} Q^o\delta a^T\Phi r dz dr - \int_s q^o + \alpha\Theta_m)\delta a^T\Phi r ds + \int_{r_1}^{r_2}\int_{-\frac{\ell}{2}}^{\frac{\ell}{2}} \rho c\delta a^T\Phi\Phi^T\dot{a} r dz dr$$

where $\dot{a}$ denotes time derivative of vector a.
The quantities α and $(q^o + \alpha T_m)$ are supposed to be invariable along each side and λ, Q^o and ρc are assumed constant over the interior of individual elements. Now the integrations affect vector Φ and its derivatives only.
Hence

$$\delta P_e = \delta a^T B a - \delta a^T L + \delta a^T H \dot{a} \tag{14}$$

where B and H represent twelve by twelve symmetric matrices computed from

$$B = \lambda\int_{r_1}^{r_2}\int_{-\frac{\ell}{2}}^{\frac{\ell}{2}} \{\Phi_z\Phi_z^T + \Phi_r\Phi_r^T\} r dz dr + \alpha \int_s \Phi\Phi^T r ds$$

and

$$H = \rho c \int_{r_1}^{r_2}\int_{-\frac{\ell}{2}}^{\frac{\ell}{2}} \Phi\Phi^T r dz dr$$

L is a twelve dimensional vector given by

$$L = Q^o \int_{r_1}^{r_2}\int_{-\frac{\ell}{2}}^{\frac{\ell}{2}} \Phi r dz dr + (q^o + \alpha T_m) \int_s \Phi r ds$$

The elemental shape allows to compute all integrals in closed form.
The relation between the degrees of freedom and coefficients a_0 thru a_{11} is found after substitution in expressions (9) and (13) of the coordinates of each of the nodal points.
Formal solution of the resulting set of twelve equations is possible if no nodal points coincide.

$$a = C\Theta \tag{15}$$

where vector Θ contains the degrees of freedom of the element, matrix C is given in Fig. 2.
Substitution in equation (14) gives

$$\delta P_e = \delta\Theta^T C^T BC\Theta - \delta\Theta^T C^T L + \delta\Theta^T C^T HC\dot{\Theta} \tag{16}$$

The sum of all elemental contributions can now be computed and may be written as

$$\delta\bar{P} = \sum_e \delta P_e = \delta\bar{\Theta}^T\{\bar{B}\,\bar{\Theta} + \bar{H}\,\dot{\bar{\Theta}} - \bar{L}\} = 0 \tag{17}$$

where $\bar{\Theta}$ denotes the vector of all degrees of freedom in the system and $\dot{\bar{\Theta}}$ means its time derivative while $\bar{B}$ and $\bar{H}$ are respectively obtained from matrix superposition - as described in chapter 3 - of C^TBC and C^THC computed for each individual element. $\bar{L}$ is derived from superposition of individual vectors C^TL.
The expression in parenthesis found in equation (17) has to equal zero since δP vanishes for all variations of the degrees of freedom

$$\bar{B}\,\bar{\Theta} + \bar{H}\,\dot{\bar{\Theta}} = \bar{L} \tag{18}$$

$\bar{B}$ is the "conductivity" matrix and $\bar{H}$ may be called the "capacity" matrix. Both are positive definite and symmetric. Equation (18)

$$
C = \begin{pmatrix}
\frac{R_2^3-3R_2^2R_1}{2e^3} & \frac{\ell R_2}{8e} & -\frac{R_2^2R_1}{2e^2} & \frac{R_2^3-3R_2^2R_1}{2e^3} & -\frac{\ell R_2}{8e} & -\frac{R_2^2R_1}{2e^2} & \frac{3R_2R_1^2-R_1^3}{2e^3} & -\frac{\ell R_1}{8e} & -\frac{R_2R_1^2}{2e^2} & \frac{3R_2R_1^2-R_1^3}{2e^3} & \frac{\ell R_1}{8e} & -\frac{R_2R_1^2}{2e^2} \\
-\frac{3R_2^3-8R_2^2R_1+R_2R_1^2}{2\ell e^3} & -\frac{R_2}{4e} & \frac{R_2^2R_1}{\ell e^2} & \frac{3R_2^3-8R_2^2R_1+R_2R_1^2}{2\ell e^3} & -\frac{R_2}{4e} & -\frac{R_2^2R_1}{\ell e^2} & \frac{R_2^2R_1-8R_2R_1^2+3R_1^3}{2\ell e^3} & \frac{R_1}{4e} & \frac{R_2R_1^2}{\ell e^2} & -\frac{R_2^2R_1-8R_2R_1^2+3R_1^3}{2\ell e^3} & \frac{R_1}{4e} & -\frac{R_2R_1^2}{\ell e^2} \\
0 & -\frac{R_2}{2\ell e} & 0 & 0 & \frac{R_2}{2\ell e} & 0 & 0 & \frac{R_1}{2\ell e} & 0 & 0 & -\frac{R_1}{2\ell e} & 0 \\
\frac{2R_2}{\ell^3 e} & \frac{R_2}{\ell^2 e} & 0 & -\frac{2R_2}{\ell^3 e} & \frac{R_2}{\ell^2 e} & 0 & -\frac{2R_1}{\ell^3 e} & -\frac{R_1}{\ell^2 e} & 0 & \frac{2R_1}{\ell^3 e} & -\frac{R_1}{\ell^2 e} & 0 \\
\frac{3R_2R_1}{e^3} & -\frac{\ell}{8e} & \frac{R_2^2+2R_2R_1}{2e^2} & \frac{3R_2R_1}{e^3} & \frac{\ell}{8e} & \frac{R_2^2+2R_2R_1}{2e^2} & -\frac{3R_2R_1}{e^3} & \frac{\ell}{8e} & \frac{2R_2R_1+R_1^2}{2e^2} & -\frac{3R_2R_1}{e^3} & -\frac{\ell}{8e} & \frac{2R_2R_1+R_1^2}{2e^2} \\
\frac{R_2^2-14R_2R_1+R_1^2}{2\ell e^3} & \frac{1}{4e} & -\frac{R_2^2+2R_2R_1}{\ell e^2} & -\frac{R_2^2-14R_2R_1+R_1^2}{2\ell e^3} & \frac{1}{4e} & \frac{R_2^2+2R_2R_1}{\ell e^2} & -\frac{R_2^2-14R_2R_1+R_1^2}{2\ell e^3} & -\frac{1}{4e} & -\frac{2R_2R_1+R_1^2}{\ell e^2} & \frac{R_2^2-14R_2R_1+R_1^2}{2\ell e^3} & -\frac{1}{4e} & \frac{2R_2R_1+R_1^2}{\ell e^2} \\
0 & \frac{1}{2\ell e} & 0 & 0 & -\frac{1}{2\ell e} & 0 & 0 & -\frac{1}{2\ell e} & 0 & 0 & \frac{1}{2\ell e} & 0 \\
-\frac{2}{\ell^3 e} & -\frac{1}{\ell^2 e} & 0 & \frac{2}{\ell^3 e} & -\frac{1}{\ell^2 e} & 0 & \frac{2}{\ell^3 e} & \frac{1}{\ell^2 e} & 0 & -\frac{2}{\ell^3 e} & \frac{1}{\ell^2 e} & 0 \\
-\frac{3(R_2+R_1)}{2e^3} & 0 & -\frac{2R_2+R_1}{2e^2} & -\frac{3(R_2+R_1)}{2e^3} & 0 & -\frac{2R_2+R_1}{2e^2} & \frac{3(R_2+R_1)}{2e^3} & 0 & -\frac{R_2+2R_1}{2e^2} & \frac{3(R_2+R_1)}{2e^3} & 0 & -\frac{R_2+2R_1}{2e^2} \\
\frac{3(R_2+R_1)}{\ell e^3} & 0 & \frac{2R_2+R_1}{\ell e^2} & -\frac{3(R_2+R_1)}{\ell e^3} & 0 & -\frac{2R_2+R_1}{\ell e^2} & -\frac{3(R_2+R_1)}{\ell e^3} & 0 & \frac{R_2+2R_1}{\ell e^2} & \frac{3(R_2+R_1)}{\ell e^3} & 0 & -\frac{R_2+2R_1}{\ell e^2} \\
\frac{1}{e^3} & 0 & \frac{1}{2e^2} & \frac{1}{e^3} & 0 & \frac{1}{2e^2} & -\frac{1}{e^3} & 0 & \frac{1}{2e^2} & -\frac{1}{e^3} & 0 & \frac{1}{2e^2} \\
\frac{-2}{\ell e^3} & 0 & -\frac{1}{\ell e^2} & \frac{2}{\ell e^3} & 0 & \frac{1}{\ell e^2} & \frac{2}{\ell e^3} & 0 & -\frac{1}{\ell e^2} & -\frac{2}{\ell e^3} & 0 & \frac{1}{\ell e^2}
\end{pmatrix}
$$

where $e=r_2-r_1$; $R_2=r_2$; $R_1=r_1$

Fig. 2: Combination matrix for Hermitian element Ring-12

represents a set of linear differential equations and solutions may be achieved by the use of a finite difference scheme. The simplest scheme is to use forward differences, but unless suitable small time increments are used the solution may become unstable (Ref. |4|). Wilson and Clough (Ref. |5|) and Zienkiewicz (Ref. |6|) suggest a stable and more accurate alternative. When steady-state conditions are involved equation (18) of course reduces to a set of linear algebraic equations

$$\overline{B}\,\overline{\Theta} = \overline{L} \tag{19}$$

which are often solved by triangular decomposition.

4.2 About the accuracy.

In Chapter 2 it has been mentioned that the basic governing equation for heat conduction in solids (Eq. (6)) can be reduced to the form

$$Au = f$$

where A is adjoint, linear and positive definite.
The solution u_o is identical to the solution of the minimization problem

$$\min_{u \in H_A} F(u) = \min_{u \in H_A} (Au,u) - 2(u,f)$$

or

$$\min_{u \in H_A} F(u) = \min_{u \in H_A} ||u-u_o||^2_A - ||u_o||^2_A \tag{20}$$

From equation (20) we see:

$$F(v) - F(u_o) = ||v-u_o||^2_A \text{ for every } v \in H_A$$

Now, if $\hat{u}$ is an approximate solution obtained from finite element analysis, then:

$$\begin{aligned} F(\hat{u}) - F(u_o) &= ||\hat{u}-u_o||^2_A \\ &= \min_{v_n \in W_n} \{F(v_n)\} - F(u_o) \\ &= \min_{v_n \in W_n} \{F(v_n) - F(u_o)\} \qquad (21) \\ &= \min_{v_n \in W_n} ||v_n-u_o||^2_A \end{aligned}$$

	r = 1		r = 2		r = 3		r = 4		r = 5	
	$\varepsilon(\Theta)$	$\varepsilon(\frac{\partial\Theta}{\partial r})$	$\varepsilon(\Theta)$	$\varepsilon(\frac{\partial\Theta}{\partial r})$	$\varepsilon(\Theta)$	$\varepsilon(\frac{\partial\Theta}{\partial r})$	$\varepsilon(\Theta)$	$\varepsilon(\frac{\partial\Theta}{\partial r})$	$\varepsilon(\Theta)$	$\varepsilon(\frac{\partial\Theta}{\partial r})$
1 Ring-12	0.	-16.1	-	-	-	-	-	-	0.	16.7
2 Ring-12	0.	- 6.4	-	-	0.54	5.0	-	-	0.	2.1
4 Ring-12	0.	- 1.8	0.07	0.48	0.03	3.0	0.07	0.13	0.	0.08
2 Ring-4	0.	-	-	-	5.0	-	-	-	0.	-
4 Ring-4	0.	-	1.3	-	1.6	-	1.8	-	0.	-

Table 1: Relative errors (in %) of data obtained with linear and third order elements

with W_n a Hamel basis of linearly independent elements. So also:

$$\min_{v_n \in W_n} ||v_n - u_o||_A^2 \leqq ||\hat{u}_o - u_o||_A^2 .$$

Here $\hat{u}_o$ is the interpolating polynomial belonging to u_o. Therefore it can be concluded that, at least in energynorm, the error introduced by the finite element method when applied to adjoint problems is smaller than or equal to the error made by interpolation of the exact solution by linearly independent elements from the Hamel basis.

From interpolation theory it is known that the error in p th derivative of a complete n th degree Hermitean interpolation polynomial is of order h^{n+1-p} where $h = \max_i ||x_i||$ is the so-called "diameter".

Furthermore it can be shown that the error contained in an incomplete polynomial is of the same order as the error caused by the largest complete polynomial which it contains. Thus the temperature approximated by the Ring-12 element implies an error of order (h^4) while computation using linear elements, e.g. Ring-4 with $\Theta = c_0 + c_1 z + c_2 r + c_3 zr$ leads to an $O(h^2)$ error. Also the error contained in the radial derivative of temperature computed with Ring-12 is of order (h^3). The difference in accuracy between the two elements may be demonstrated on the basis of the problem of purely radial heat conduction in a thickwalled cylinder.

With an outer to inner diameter ratio equal to 5 and a 100^oC temperature difference over the wall thickness the temperature profile is given by

$$\Theta = \frac{100}{\ln 5} (1 - \ln r)$$

where r means radius divided by inner radius.

Numerical results have been calculated for equidistant divisions of the wall thickness in two and four elements of both types. Table 1 shows that the relative errors $\varepsilon(\Theta)$ and $\varepsilon(\frac{\partial\Theta}{\partial r})$ computed from these results are within the limits of the error extimates. The absolute deviations of the temperature profile interpolated in accordance with the appropriate polynomial are plotted in Fig. 3 . From this we observe good agreement of accuracy between one Hermitean element and four linear elements. This will hold also when axial heat conduction is involved and therefore in the case of general two-dimensional distributions one fourth order element will give about the same accuracy as sixteen linear elements. For many practical applications however the accuracy required will not justify the choice of Ring-12 elements and idealization scheme's on the basis of the Ring-4 elements will lead to satisfying results. The use of Hermitean elements for temperature calculations is hampered if materials are involved with different values of the coefficient of heat conduction. At an interface of two such materials the continuity of heat flux means a discontinuity of the temperature derivative normal to the interface. Therefore degrees of freedom at interfaces consisting of temperature derivatives are undefined if no special provision is made.

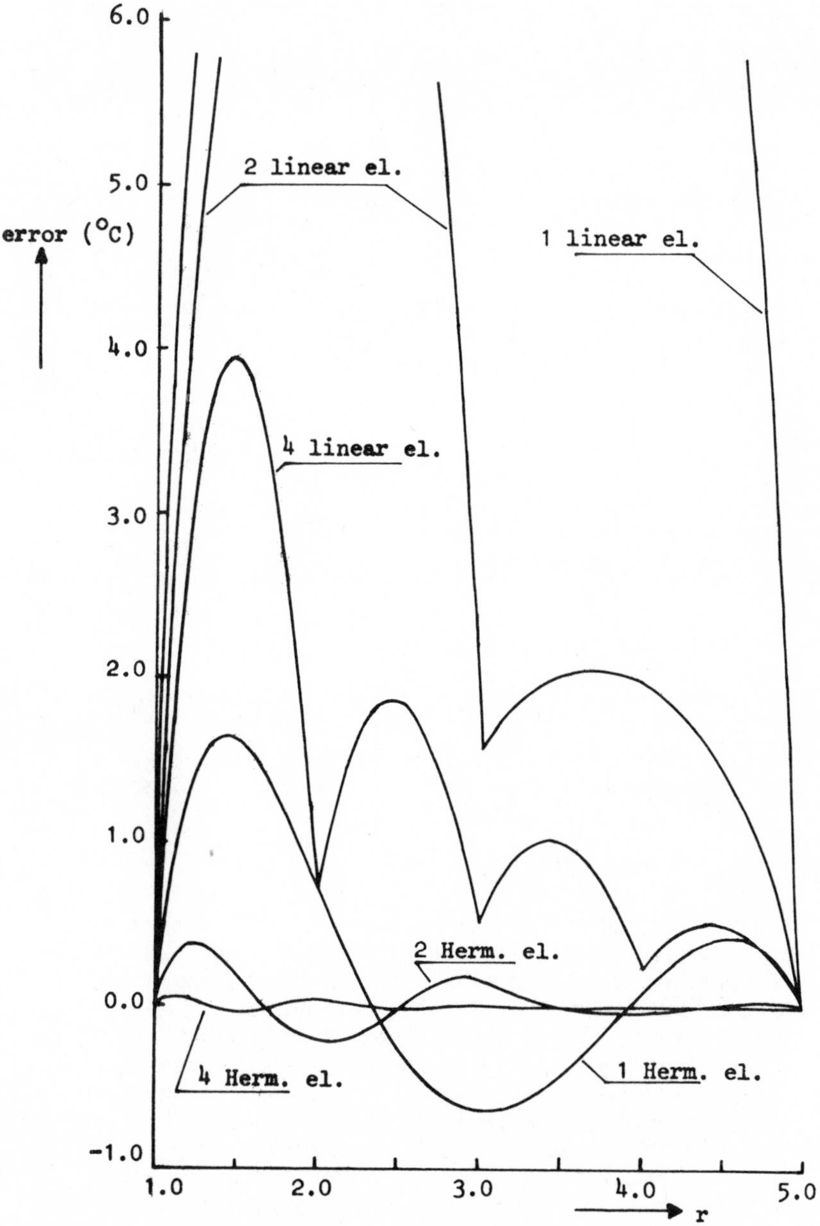

Fig. 3: Deviations of calculated and interpolated results

4.3 An application.

The FIESTA code has been used to determine the steady state temperature distribution in a pump housing. The pump, which is of the centrifugal type, is provided with three concentric channels serving for cooling and two concentric channels, which are filled with insulating material and situated between the fan and the cooling channels. At the inner side, the housing of the fan is furnished with axial-wearing surfaces. The external surface of the pump is insulated.

The range of parameters appears as follows:

Pumped medium temperature	T_{m1}	$= 250^{\circ}C$
Cooling water temperature	T_{m2}	$= 55^{\circ}C$
Coefficient of heat transfer (α), between coolant and housing for each of the three cooling channels in sequence of increasing radius	α_1	$= 0.107\ W\ cm^{-2}\ {}^{o}C^{-1}$
	α_2	$= 0.101\ W\ cm^{-2}\ {}^{o}C^{-1}$
	α_3	$= 0.099\ W\ cm^{-2}\ {}^{o}C^{-1}$
α between pumped medium and structure inside the inlet channel of the pump	α_4	$= 0.504\ W\ cm^{-2}\ {}^{o}C^{-1}$
α between pumped medium and structure inside the fan housing, α_5 varies with radius from		$0.77\ W\ cm^{-2}\ {}^{o}C^{-1}$
	until	$1.79\ W\ cm^{-2}\ {}^{o}C^{-1}$
Housing coefficient of heat conductivity		$0.146\ W\ cm^{-1}\ {}^{o}C^{-1}$
Wearing surface coefficient of heat conductivity		$0.291\ W\ cm^{-1}\ {}^{o}C^{-1}$

Fig. 4 shows a longitudinal section of the pump housing with dimensions and isothermal curves derived from temperature analysis.

5. THE NASTRAN COMPUTER PROGRAM.

There is a large degree of similarity between the finite element formulation of problems from structural mechanics and temperature analysis. Comparison of the expression derived from variation of the potential energy with equation (4) learns that temperature can be substituted for one of the displacement components, heat fluxes take the place of stresses and heat capacity substitutes structural mass. The heat conduction matrix for a volume heat conduction element may be derived from a thermal potential function in the same way that the stiffness matrix of a structural element is derived from the strain energy function. The thermal potential function is defined by the first term of the expression F(u) (Eq. (8)).

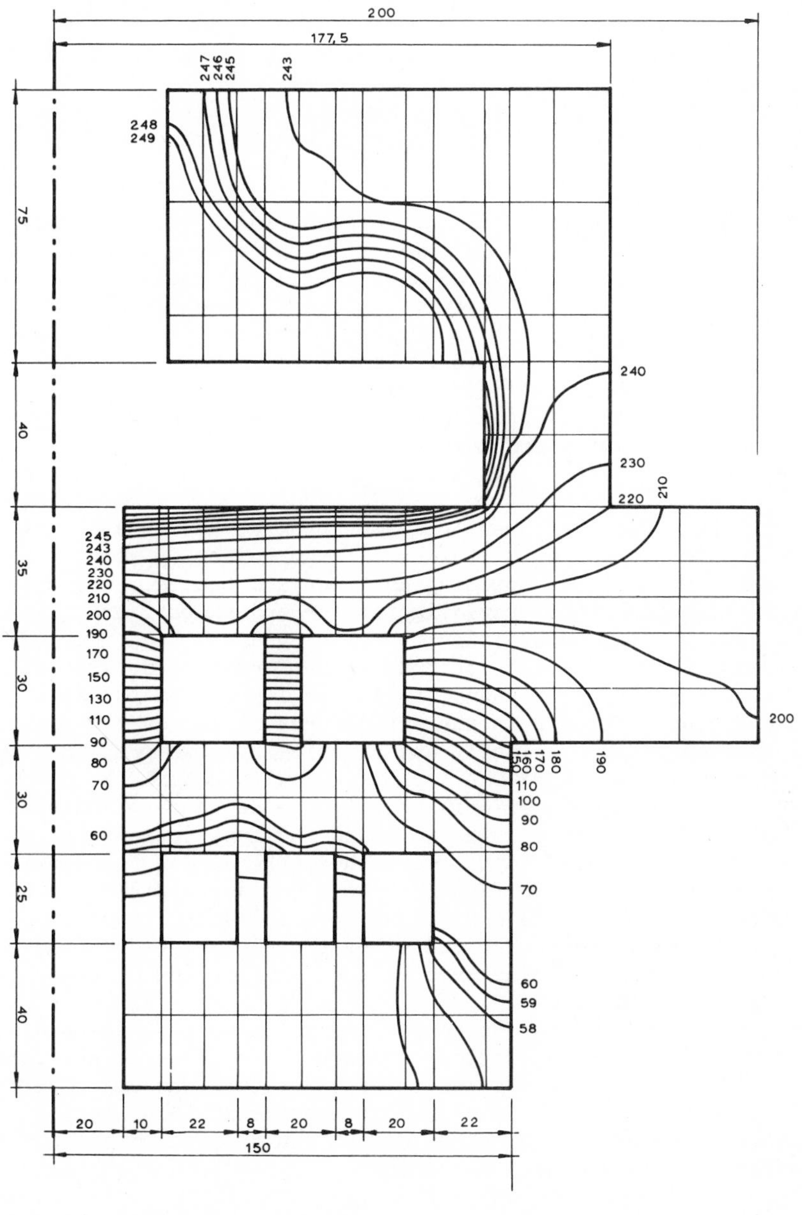

Fig. 4: Longitudinal section of pump housing with isothermal curves

$$U = (Au,u) = - \int q.\nabla u \, dV$$

where q is the heat flux density, ∇u is the temperature gradient, and the integration is performed over the volume V. The main deviations exist in the constitutive equation and the boundary conditions relative to the two phenomena.
Hooke's law deals with vectorial quantities and it is essentially more complicated than Fourier's law of heat conduction. Therefore the stiffness matrix is assembled from more contributions than the heat conduction matrix.
The boundary condition defining the heat input relative to the difference of structural and ambient temperatures does not have a counterpart in the structural displacement problem. This restriction may however be overcome by introduction of so-called "convection boundary elements" with a prescribed displacement to simulate ambient temperature and a value of the coefficient of heat conduction replacing the coefficient of heat transfer.
A computer program originally developped to solve problems from structural mechanics and augmented to allow computation of temperature profiles is NASTRAN.
This acronym is formed from NASA Structural Analysis
NASTRAN is a finite element computer program designed to solve a broad class of problems. It uses a large number of subroutines assembled to modules composing matrices and tables, executing matrix operations and providing for in- and output. The last level released is 15.1.0 containing twelve programmed sequences of ordered matrix operations termed "rigid formats" for static, elastic stability and dynamic analyses. It is also possible to do cases not provided for in the rigid formats. The user can introduce modules generated by himself and organize his own problem steps by using a language called DMAP, contained within NASTRAN.
Using DMAP the user need only specify the sequence of matrix operations, the module selections and the instructions for problem flow organization needed to solve the mathematical formulation of his particular case. In fact DMAP need not necessarily pertain to structures; it can handle problems in any discipline so long as it is formulated in matrices. The other principle option in using DMAP called the ALTER feature, is the modification or augmentation of a given rigid format.
The adaption of NASTRAN to handle thermal problems has been accomplished after convection boundary elements were incorporated and the module assembling the stiffness matrix was adjusted to enable generation of the heat conduction matrix. Thus stationary and transient temperature problems can be analyzed using rigid formats available for structural mechanics.

5.1 An example.

It has been mentioned that many properties can be described by

special forms of the general equation of balance (Eq. (1)). The application of Galerkin's method for such properties is straight forward and the finite element formulation may be obtained without much difficulties. A phenomenon not described by an equation derived from equation (1) is the diffusion of neutrons. In particular when the neutrons belong to only one energy group this phenomenon shows a close similarity with heat transfer in solids and Galerkin's method may be applied in exactly the same way to arrive at a starting point for the finite element approach.
The equation of one-group neutron diffusion reads (Ref. |8|)

$$-\frac{1}{v}\frac{\partial\psi}{\partial t} + D\frac{\partial^2\psi}{\partial x_i \partial x_i} - \sigma\psi + S = 0 \qquad (22)$$

Here v stands for velocity, D for the diffusion coefficient, ψ for the scalar flux, σ for the macroscopic absorption cross section, and source term S for the rate of production of neutrons per unit volume per second.

The boundary conditions applied to equation (22) involve continuity of flux and current at interfaces, and vanishing of flux at extrapolated boundaries or return current at the actual boundaries.
The general form of the boundary conditions reads

$$D\frac{\partial\psi}{\partial n} + \gamma\psi + q = 0 \qquad (23)$$

where γ and q are constants. Equations (22) and (23) can be combined to obtain the functional:

$$H = \iiint\limits_V \left(\frac{1}{v}\frac{\partial\psi}{\partial t}\psi + \frac{D}{2}\frac{\partial\psi}{\partial x_i}\frac{\partial\psi}{\partial x_i} + \frac{\sigma}{2}\psi^2 - S\psi\right)dV + \iint\limits_A \left(q\psi + \frac{\gamma}{2}\psi^2\right)dA \qquad (24)$$

Functional H reaches a minimum value with respect to all admissible variations of flux distribution ψ. Comparison of expressions (5) and (24) shows that duality of finite-element temperature programs is obtained when for neutron diffusion the following substitutions are made: $\frac{1}{v}$ for ρc; D for λ; S for Q^o; $-\frac{q}{\gamma}$ for Θ_m; and γ for α.

Moreover, the heat capacity matrix has to be generated using a coefficient of thermal capacity equal to σ and added to the heat conduction matrix.
For static analyses the relevant modifications to the existing rigid format can be introduced by addition of the input data cards given in Table 2.
The method described has been used to determine the flux distribution in the basic area of symmetry in a fuel moderator mixture located between a square array of cruciform control rods. The rods are black to thermal neutrons and a uniform slowing density has

been assumed. Figure 5 shows the geometry together with the idealization and calculated lines of constant neutron flux.

```
ALTER 22,22
$ REMOVE STATEMENT 22 FROM THE DMAP SEQUENCE FOR STATIC ANALYSIS
$ AND REPLACE IT BY:
PARAM //C,N,NOP/V,N,SKPMGG=1 $
$ SKPMGG=1 WILL ENABLE COMPUTATION OF HEAT CAPACITY MATRIX MGG
ALTER 37
$ AFTER DMAP STATEMENT 37 INSERT:
ADD KGGX,MGG/KGG1/C,Y,ALPHA=(1.0,0.0)/C,Y,BETA=(1.0,0.0) $ AND
CHKPNT KGG1 $
$ ADDITION OF HEAT CONDUCTION AND HEAT CAPACITY MATRICES GIVES KGG1
ALTER 44,44
$ REPLACE STATEMENT 44 BY:
EQUIV KGG1,KGG/NOGENL $
$ KGG=KGG1; KGG IS RECOGNIZED BY SUBSEQUENT MODULES
ENDALTER
```

Table 2: ALTER package for neutron diffusion calculations

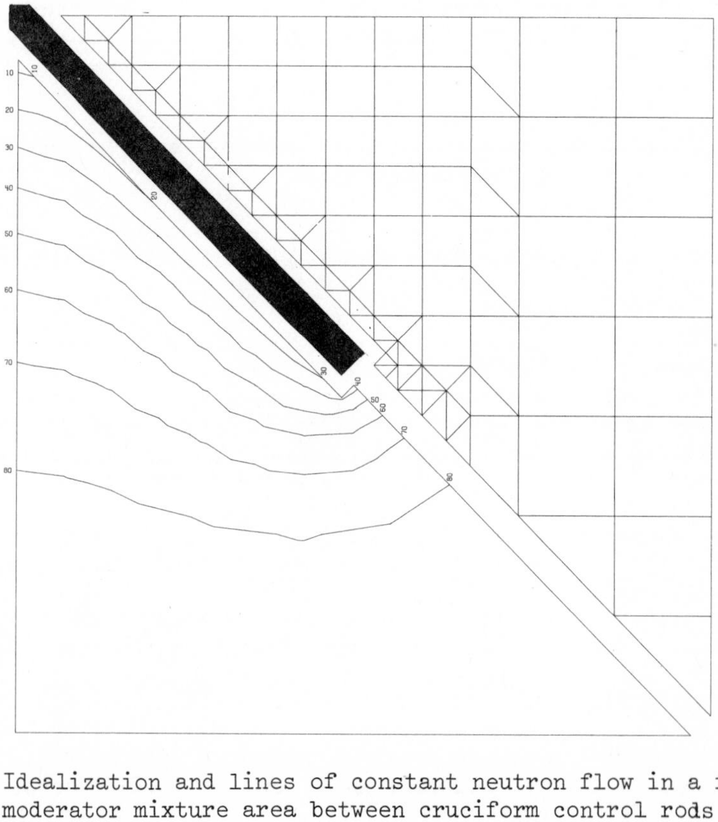

Fig. 5: Idealization and lines of constant neutron flow in a fuel-moderator mixture area between cruciform control rods

LITERATURE.

1. Visser, W. The finite element method in deformation and heat conduction problems, thesis, Delft, 1968.

2. Oliviera, E.R.A. Int. J. Solids Structures 4 (1968) 929.

3. Patterson, C. Sufficient conditions for convergence in the finite element method for any solution of finite energy, Sheffield 1972.

4. Grandall, S.H. Engineering Analysis, London, Mc Graw-Hill Book Cy., 1956.

5. Wilson, E.L. and R.W. Clough Symp. on the use of computers in civil engineering, Lisbon, 1962.

6. Zienkiewicz, O.C. The finite element method in engineering science, London, Mc Graw-Hill Book Cy., 1971.

7. Patterson, C. "A class of functionals giving improved convergence with finite elements", International conference on Variational Methods in Engineering, University of Southampton (1972).

8. Weinberg, A.M. and E.P. Wigner The physical theory of neutron chain reactors, Chicago, The University of Chicago Press, 1958.

9. Zienkiewicz, O.C. Appl. Mech. Rev. 23 (1970) 249.

10. Elsgolc, L.E. Calculus of Variations, London, Pergamon Press, 1969.

11. Przemieniecki, J.S. Theory of matrix structural analysis, London, Mc Graw-Hill Book Cy., 1968.

12. Lanczos, C. The variational principles of mechanics, Toronto, University of Toronto Press, 1966.

13. Zlámal, M. Numer. Math. 12 (1968) 394.

14. Oden, J.T. Int. J. Num. Meth. Eng. 1 (1969) 205, 247.

15. Oden, J.T. Nucl. Eng. and Design 10 (1969) 465.

16. Schechter, R.S. The variational method in engineering, London, Mc Graw-Hill Book Cy., 1967.

17. Crastan, V. Nucl. Eng. and Design 15 (1971) 54.

18. Argrys, J.H and D.W: Scharpf Nucl. Eng. and Design 10 (1969) 456.

19. Emery, A.F. and W.W. Carson J. of Heat Transfer 93 (1971) 136.

20. Babuška, I. SIAM J. on Num. An. 8 (1971) 304.

21. Finlayson, B.A. SIAM J. on Num.An. 8 (1971) 316.

22. Mac Neal, R.H., editor: The NASTRAN Theoretical Manual. NASA SP-221(01) (1972).

23. McCormick, C.W., editor: The NASTRAN User's Manual. NASA SP-222(01) (1972).

LIST OF PARTICIPANTS

NAME	ADDRESS (at the time of the meeting)
Arkuszewski, J.	Computing Center Cryfronet, IBJ, Swierk 05400 Otwock, Poland
Barker, V.A.	Numerisk Institutt, The Technical University of Denmark, 2800 Lyngby Denmark
Barrett, K.E.	Lanchester Polytechnic, Dept. of Mathematics and Statistics, Priory Street, Coventry CV1 5FB, UK
Bech, N.	AEK Risø Research Esablishment, 4000 Roskilde. Denmark
Bergan, P.G.	Inst. for Statikk, NTH, University of Trondheim, 7034 Trondheim, Norway
Boonstra, B.H.	Reactor Centrum Nederland, Petten (NH), The Netherlands
Borysiewicz, M.	Computing Center Cyfronet, IBJ, Swierk 05400 Otwock, Poland
Bø, K.	RUNIT, NTH, University of Trondheim, 7034 Trondheim, Norway
Carmignani, B.	Ist. Impianti Nucleari fac. Ingegneria, Universita di Pisa, Italy
Chenin, P.	Université de Grenoble, B.P. 53, Centre de Tri, 38061 Grenoble-Cedex, France
Collatz, L.	Inst. für Angewandte Mathematik, Universität Hamburg, Rothenbaumchausse 67/69, Hamburg 13, Germany

Daneri, A.	Centro di Calcolo, CNEN, Via Mazzini 2, 40 138 Bologna, Italy
Davierwalla, D.	Swiss Federal Institute for Reactor Research, 5303 Würenlingen, Switzerland
Debler, W.	Dept. of Applied Mechanics and Engineering Science, University of Michigan, Ann Arbor, Michigan 48104, USA
Engquist, B.	Dept. of Computer Sciences, Uppsala University, Box 256, 751 05 Uppsala, Sweden
Finkelstein, L.	Israel Atomic Energy Commission, Soreq Nuclear Research Centre, Yavne, Israel
Fladmark, G.E.	Institutt for Atomenergi, P.O.B. 40, 2007 Kjeller, Norway
Gardner, L.R.	School of Mathematics, UCNW, University of Wales, Bangor, Caerns, UK
Gautier, A.	G.A.A.A., 20 Avenue Edouard Herriot, 92-Le Plessis Robinson, France
Geertsma, J.	Koninklijke/Shell Exploratie & Produktie Laboratorium, Volmerlaan 6, Rijksvijk (ZH), The Netherlands
George, A.	Dept. of Applied Analysis & Computer Science, University of Waterloo, Waterloo, Ontario N2L 3G1, Canada
Gourlay, A.R.	Dept. of Mathematics, Loughborough University, Loughborough, Leicestershire, UK
Gregory, R.T.	Center of Numerical Analysis, The University of Texas, Austin, Texas 78712, USA
Hansteen, O.E.	Norwegian Geotechnical Institute, P.O.B. 40, Tåsen, Oslo 8, Norway
Hoppe, V.H.	Burmeister & Wain, Torvegade 2, 1449 Copenhagen, Denmark
Holand, I.	Inst. for Statikk, NTH, University of Trondheim, 7034 Trondheim, Norway
Ingham, D.B.	Dept. of Applied Mathematical Studies, The University, Leeds LS2 9JT UK

Klumpp, W.	Inst. für Kernenergetikk der Universität Stuttgart, Pfaffenwaldring 31, 7 Stuttgart 80, Germany
Kinnebrock, W.	Gesellschaft für Kernforschung, Postfach 3640, 75 Karlsruhe, Germany
Kulikowska, T.	Computing Center Cyfronet, IBJ, Swierk 05400 Otwock, Poland
Lang Rasmussen, O.	AEK Risø Research Establishment, 4000 Roskilde Denmark
Lemanska, M.	Israel Atomic Energy Commission, Soreq Nuclear Centre, Yavne, Israel
Lezius, D.K.	NASA-Ames Research Center, Moffett Field, Cal. 94035, USA
Megård, G.	The Norwegian Building Research Institute, Forskningsvn. 3b, Oslo 3, Norway
Mitchell, A.R.	Dept. of Mathematics, The University, Dundee DD1 4HN, UK
Moen, H.	Institutt for Atomenergi, P.O.B. 40, 2007 Kjeller, Norway
Moshagen, H.	SINTEF, NTH, University of Trondheim, 7034 Trondheim, Norway
Nerlich, K.D.	Technischer Überwachungs-Verein Bayern E.V., Landsbergerstr. 182, 8 München 21, Germany
Prij, J.	Reactor Centrum Nederland, Petten (NH), The Netherlands
Rall, L.B.	University of Wisconsin, USA, p.t. Oxford University Computing Lab., 19 Parks Road, Oxford OX1 3PL, UK
Rasmussen, H.	Lab. for Applied Mathematical Physics, The Technical University of Denmark, 2800 Lyngby, Denmark
Reefman, R.B.J.	Hazemeyer BV, P.O.B. 23, Hengelo (O), The Netherlands
Rosser, J. Barkley	Mathematics Dept., Brunel University, Uxbridge, Middlesex, UK
Samuelsson, A.	Dept. of Structural Mechanics, Chalmers Tekn. Högskola, Göteborg 5, Sweden

Saydam, T.	Istanbul Technical University, Spor Cad. 138/6 Macka, Istanbul, Turkey
Schmidt, F.A.R.	Inst. für Kernenergetik der Universität Stuttgart, Phaffenwaldring 31, 7 Stuttgart 80, Germany
Schnierle, H.	Technischer Überwachungs-Verein Bayern E.V., Eichstätter Str. 5, 8 München 21, Germany
Siemieniuch, J.L.	Dept. of Mathematics, University of Manchester, Oxford Road, Manchester, UK
Sigurdsson, S.T.	Raunvisindastofnun Haskolans, Dunhaga 3, Reykjavik, Iceland
Skappel, J.	Norwegian Computing Centre, Forskningsvn. 1b, Oslo 3, Norway
Skordalakis, E.	Greek Atomic Energy Commission, N.R.C. "Democritos", Computer Center, Aghia Paraskevi-Attikis, Athens, Greece
Smoldern, J.	von Karman Institute for Fluid Dynamics, Chausée de Waterloo 72, B-1640 Rhode-St-Genese, Belgium
Spreeuw, E.	Reactor Centrum Nederland, Petten (NH), The Netherlands
Stankiewicz, R.	Computing Center Cyfronet, IBJ, Swierk 05400 Otwock, Poland
Sundström, A.	The Research Institute of the Swedish National Defence, 104 50 Stockholm 80, Sweden
Sæbø, T.	A/S Årdal og Sunndal Verk, 5875 Årdalstangen, Norway
Toselli, G.C.	Centro di Calcolo, CNEN, Via Mazzini 2, 40 138 Bologna, Italy
Trulsen, J.	Auroral Observatory, University of Tromsø, 9000 Tromsø, Norway
Ursin, B.	Inst. for teknisk kybernetikk, NTH, University of Trondheim, 7034 Trondheim, Norway
Varga, R.S.	Dept. of Mathematics, Kent State University, Kent, Ohio 44242, USA
Versluis, R.M.	OECD Halden Reactor Project, P.O.B. 173, 1751 Halden, Norway

Weiss, Z.	AB ASEA-ATOM, KCD, Box 53, 72 104 Västerås 1, Sweden
Whiteman, J.R.	Dept. of Mathematics, Brunel University, Uxbridge, Middlesex, UK
Winther, R.	University of Oslo, Blindern, Oslo 3, Norway
Wirz, H.J.	von Karman Institute for Fluid Dynamics , Chaussée de Waterloo 72, B-1640 Rhode-St-Genese, Belgium
Young, D.M. Jr.	Center of Numerical Analysis, The University of Texas, Austin, Texas 78712, USA
Zienkiewics, O.G.	Dept. of Civil Engineering, University of Wales, Singleton Park, Swansea, SA2 8PP, UK
Aasen, J.O.	Inst. of Mathematics, NTH, University of Trondheim, 7034 Trondheim, Norway

Section Secretaries

Holthe, O.
Madsen, E.
Rasmussen, L.
Fantoft, S.
Danielsen, T.

Institutt for Atomenergi, P.O.B. 40, 2007 Kjeller, Norway